125 Years
The Physical Society and
The Institute of Physics

125 Years
The Physical Society and
The Institute of Physics

Edited by
John L Lewis OBE, FInstP

With a Preface by
Sir Gareth Roberts FRS, FInstP
President of the Institute of Physics

Institute of Physics Publishing
Bristol and Philadelphia

© IOP Publishing Ltd 1999

British Library Cataloguing-in-Publication Data

A catalogue record for this book is available from the British Library.

ISBN 0 7503 0609 2

Library of Congress Cataloging-in-Publication Data are available

Published by Institute of Physics Publishing, wholly owned by The Institute of Physics, London

Institute of Physics Publishing, Dirac House, Temple Back, Bristol BS1 6BE, UK

US Office: Institute of Physics Publishing, The Public Ledger Building, Suite 1035, 150 South Independence Mall West, Philadelphia, PA 19106, USA

Typeset in TeX using the IOP Bookmaker Macros
Printed in the UK by J W Arrowsmith Ltd, Bristol

The Institute of Physics was granted armorial bearings by the Kings of Arms on 9 June 1993.

The Institute's coat of arms is described in heraldic language as follows:

> Shield—vert; a representation of a nuclear atom with three electron orbits or; a chief wavy of four waves; a central trough enhanced ermine; and, for the crest, upon a helm with a wreath argent and vert between two sprigs of oak

vert, a thunderbolt or; enflamed proper mantelled vert doubled argent.

Supporters—on either side a stag proper, attired and unguled or; gorged with a wreath argent and vert; the compartment comprising a grassy mount proper.

The design is intended to reflect the wide range of interests and activities of the Institute. The predominant colour of the shield and other parts of the arms is green, as in the natural environment. The components of the shield reflect symbolically some of the main themes of physics: atoms through the central symbol, waves through the single-slit diffraction pattern and fields through the ermine. The helm (helmet), situated directly above the shield, has a closed visor representing the fact that the Institute is a corporate body. The crest—a winged, flamed thunderbolt between oak leaves—is joined to the helm by twisted ribbons coloured alternately green and silver and has a double reference: the flamed thunderbolt symbolises electrical and thermal energy and the oak leaves have been taken from the tree of knowledge, used as part of the badge of the Physical Society. The stag supporters are inspired by the stag's head that appears on the shield of Lord Kelvin (President of the Physical Society, 1880–1882).

The motto—*Intellegite et Explicate*—may be translated as 'understand and explain', drawing attention to the Institute's aim of understanding the physical universe and explaining it to the world. It reflects the Royal Charter granted in 1970 which states: 'The object for which the Institute is hereby constituted is to promote the advancement and dissemination of education in the science of physics, pure and applied'.

CONTENTS

PREFACE

I am delighted to be President of the Institute of Physics during the closing years of the 20th century when it is appropriate to look back at the development of the Institute and its forerunner, the Physical Society, formed in 1874. Physics has had a glorious 125 years, the prospects for the future are as bright as they have ever been, and there is no more appropriate time to reflect.

The late-19th and 20th centuries have revealed to us riches of the physical world that the founder members of the Physical Society could not possibly have envisaged. Developments based on a constant stream of discoveries by great physicists of the past and present have led to the physics of today and the rich tapestry of physics-based industries that contribute so much to our well-being. In addition, physics has become the enabling science for major innovations in biology, medicine and engineering.

This book is an account of the evolution of the Physical Society, the learned society for physics in the United Kingdom and Ireland, from its origins in 1874 through its merger in 1960 with the professional body for physicists, the Institute of Physics, up to the present day. It is a story of physicists with vision, determination and concern for their subject who ensured that physics was well served and that their fellow physicists had an organisation to turn to for advice and support.

During this period the Institute has become a world leader in publishing—a quintessential component of all learned societies' activities. Electronic publishing, which was not realisable until very recently and which would have been an utterly foreign concept to the Institute's forefathers, has been developed so successfully by the Institute that others from around the world turn to our publishing

Sir Gareth Roberts, President of the Institute of Physics.

organisation to experience the state-of-the-art technology. In addition, the Institute has an enviable reputation for organising conferences, and manages a portfolio of activities to support physics education in schools, colleges and universities that is of the highest order and of fundamental relevance to the future of our subject. Last, but not least, the Institute has increasingly become the campaigning voice of the physics community, ensuring that government and its agencies are aware of the concerns and requirements of physics and physicists.

The Institute has now become a truly international organisation with members throughout the world. Past Presidents and Nobel Laureates such as Ernest Rutherford, William Bragg and his son Lawrence, J J Thomson and his son G P, and others involved in the pioneering days at the end of the 19th and the beginning of the 20th centuries, would be astounded to learn that their Society had expanded to such an extent that in 1998 it had over 22,500 physicists in membership—either as students, associates, graduates or full members. The Institute has grown alongside the discipline of physics.

It is a great honour to be the President of the body that represents my chosen profession, but it is humbling to read the long list of distinguished physicists who have preceded me. It is an inspiration to all of us to learn, in this clearly written and well researched history, of their dedication in the early days of the Physical Society and the Institute of Physics. But for their efforts, we would not today have a strong, forward-looking and dynamic Institute that is well equipped to provide support for physics well into the 21st century. Let the next 125 years be as successful for physics and the Institute as the first 125 years have been.

Sir Gareth Roberts FRS, FInstP
President
November 1998

ACKNOWLEDGMENTS

This book owes a great deal to many people for the contributions they have made, but a particularly warm tribute is due to Dr Maurice Ebison, one-time Deputy Chief Executive of the Institute, and Mr Ron White, one-time Membership Records Manager and a regular writer on historical topics.

Other major contributions were made by Dr Kurt Paulus, a director of the Institute's publishing company, Dr Brian Manley, the President of the Institute from 1996 to 1998, Mr Arlie Bailey, one-time Vice-President for Engineering, and Dr Eric Duckworth, Director of Fulmer Research Institute from 1969 until 1990. Further help was provided by Dr Derek Jefferies, Mr Clive Jones and Mr Stephen Sadler. There were also a number of former Officers and members of the Institute who went to much trouble reading early versions and making helpful suggestions. We are most grateful for all their help.

Our thanks also go to the staff of Institute of Physics Publishing, in particular Margaret O'Gorman (Books Publisher), Sharon Toop (Production Manager) and Al Troyano (Senior Production Editor), for their assistance in bringing this book to publication.

Every effort has been made to ensure the accuracy of what has been written, but any errors must be the responsibility of the Editor and not of those listed above.

John L Lewis
Editor
November 1998

CHAPTER 1

THE EMERGENCE OF PHYSICS

The works of Aristotle were brought to England in the 13th century, but it was Francis Bacon, the Franciscan friar (c 1210–1292) who alone among his contemporaries saw the importance of experimental methods. 'There are two methods of investigation', he wrote, 'through argument and through experiment. Argument does not suffice, but experiment does.' But it needed the Renaissance to stimulate the enquiring mind which was to lead eventually to the physics of today. In the 16th century Copernicus and Galileo set the stage in Europe. By the 17th century there were entrepreneurs such as Robert Boyle, who went to Eton to study at the age of eight and had his own laboratory in Oxford at the age of twenty-seven. There he began his investigations with the aid of an Otto von Guericke pump on the elastic properties of the atmosphere, which, of course, led to Boyle's Law.

The latter part of the 17th century was dominated by Isaac Newton. It has been said:

> 'The essential strength of physics lies in the great depth of its conceptual schemes in the relatively few principles that serve to unify the entire range of man's knowledge of the physical universe'.

This is typified by Newton's laws of motion and his work on gravitation. On the death of Robert Hooke in 1703, Newton became the President of the Royal Society, to which post he was re-elected annually until his death in 1727.

In the 18th century, Coulomb (in France) was building his torsion balance to study the forces between electric charges and Benjamin

Isaac Newton.

Franklin (in America) was proposing the 'one-fluid theory of electricity'. In England, Henry Cavendish (1731–1810) was perfecting his torsion balance in order to measure the gravitational constant. Thomas Young was born in 1773 and could read at the age of two. He began studying Latin at the age of six and at sixteen had a thorough knowledge of both Latin and Greek, and also spoke eight European languages. At twenty he read a paper to the Royal Society in which he attributed the accommodation of the eye to its muscular structure and at twenty-one he became a Fellow of the Royal Society. He went on to develop his wave theory of light.

At the turn into the 19th century Augustine Fresnel (1788–1827) (in France) and Hans Christian Oersted (1777–1851) (in Denmark) were making their contributions and Michael Faraday (1791–1867) was entering the scene in England. The Royal Institution was founded by Benjamin Thompson, later Lord Rumford, in 1799 and it was here from 1801 onwards that Humphrey Davy attracted fashionable London to hear his lectures and see his demonstrations on heat and light. But at that time there was a very wide gap between the wealthy and the poor. The leisured classes could enjoy cultural activities at places such as Spring Gardens in Vauxhall where 'everywhere there is music'. On the other hand, children of four and five years old were being used as chimney sweeps in the early 1800s, even though this contravened the law that no child should be employed under nine years of age—or for more than eight hours a day.

The advancement of science during the 18th century depended almost exclusively upon the activities of the leisured classes and philosophical clubs such as the Lunar Society of Birmingham and the Literary and Philosophical Society of Manchester. Progress for the most part relied upon the activities of fact-finding amateurs, though Newton did publish his *Opticks* in 1704. Today universities play an overwhelmingly important part in the contemporary intellectual scene, but the intellectual revolution in science which had begun in the 17th century and had consequential effects throughout the 18th century developed almost entirely outside the universities.

At the beginning of the 19th century the Universities of Oxford and Cambridge retained their close links with the Church of England and scientific chairs were often given to clergymen who did not normally regard science as one of their most important activities. These universities were therefore not in a good position to deal with the first stirrings of a fear that was to become much more insistent later in the century, namely that France and Germany were becoming more proficient in harnessing their increasing scientific knowledge to industry. The industrial supremacy that England was enjoying in the middle years of the century was looking increasingly vulnerable.

The founding of the University of London in 1826 was a turning point which enabled science to become established, if initially somewhat precariously, within the institutions of higher education. It was a frail growth which fortunately soon received a stimulus from the British Association for the Advancement of Science, which held its first

Watercolour painting by Harriet Jane Moore of Michael Faraday in his basement laboratory at the Royal Institution, 1852 (reproduced by kind permission of the Royal Institution).

meeting in 1831 and from then on campaigned vigorously on behalf of science education.

In view of the contemporary importance of practical work, it may be surprising that scientific laboratories played little part in research or education prior to 1800. Anyone seeking to do research work within the physical sciences had to depend on private facilities for their investigations. Since there were no institutions where students could learn practical methods, they had to rely on working with masters as disciples or apprentices. There was certainly no pressure for the establishment of self-regulating groups in either physics or chemistry in which people could be recognised as being professionally qualified.

In Britain chemistry emerged earlier than physics as a separate discipline. The premises for the Royal Institution in Albemarle Street were completed in March 1801 and Sir Humphrey Davy became Professor of Chemistry and Michael Faraday was appointed Assistant there in 1813. Sir Humphrey died in June 1829 and Faraday became Director. The laboratory at the Royal Institution was essentially a straight chemical laboratory initially, but in due course became the first British laboratory in which experimental research in physics was done. The first university chemical laboratories were founded around 1829 in Glasgow University and University College London. By 1845 most British institutions of higher education had such chemical laboratories, but, as with Faraday's chemical laboratory at the Royal Institution, the research in those laboratories included many topics which would later be considered physics.

Further private and non-academic physics laboratories in Britain appeared in the 1830s and 1840s: Wheatstone's laboratory at King's College London in 1835, Joule's in Manchester in 1838, Forbes' in Edinburgh in 1840 and Stokes' in Cambridge in 1849. These laboratories were limited in size and relaxed in organisation. Some allowed students to do experimental work, but others permitted students only to watch demonstrations. At this time there were no requirements within degree regulations for students to undertake any practical work. If students entered a laboratory it was purely as a result of a voluntary arrangement with the laboratory director, who was of course responsible for finding the resources to run the laboratory until such time as it was formally recognised as a part of the university.

William Thomson (later Lord Kelvin) set up the first example of any sort of university physics laboratory in 1846 at Glasgow College, though without any official university recognition. The beginnings were very humble with Thomson having to manage with old apparatus about which he commented that most of it was 'worm-eaten mahogany'. Four years later Thomson was allowed to use an abandoned wine cellar near the physics lecture theatre, but official recognition of the laboratory by the College was not given until 1866, twenty years after it came into existence.

The University of London played a vital role in the movement to increase the prestige of science studies. In 1858 a major development occurred when a new charter was issued which allowed the University to award science degrees for the first time, and in 1860 the first honours BSc degree was awarded. It is interesting to note that even at this date the word physics did not occur in BSc regulations.

The regulations did include such topics as mechanical and natural philosophy; chemistry, theoretical and practical, inorganic and organic; animal physiology; geology and palaeontology; logic and moral philosophy. The development of this London BSc degree necessitated the establishment of physics laboratories for teaching students at University College in 1866 and at King's College in 1868. It would seem that King's College was probably the first to require laboratory work in physics. Laboratories were started in Oxford University in 1866, in Edinburgh University in 1868, in Owen's College in Manchester in 1870 and in Cambridge University in 1871. Of particular importance for the future of the Institute of Physics was the laboratory set up by Frederick Guthrie at the Royal School of Mines in South Kensington in 1872. Although physics was some forty years behind chemistry in the development of undergraduate teaching laboratories, by the early 1870s laboratories had become an integral part of the study of physics.

This rapid expansion in the number of university physics laboratories over a relatively short period of time owed much to the growth in applied research in thermodynamics and in electric telegraphy. Perhaps most influential was the need for the country to do something about science education following the alarm generated by the Paris Exhibition of 1867 where it had become obvious that Britain no longer occupied the dominant position in industry that had been so evident at the Great Exhibition in London in 1851. Certainly much remained to be done to move towards recognisable professionalism in physics, but at least by the 1870s it was at last possible for a student to obtain an academic training in physics, though actual degrees in physics were relatively uncommon before the 1890s.

The term 'physicist'

The process by which physics emerged as a distinct discipline from the general scientific background had a relatively late beginning in the 19th century. The topics which eventually ended up within the boundaries of physics were those that remained after astronomy, chemistry and geology had been hived off from natural philosophy. What remained were areas such as mechanics, heat, light, sound, magnetism and electricity.

It may be interesting to look briefly at the introduction of the words *physicist* and *scientist*. The former word had been used in earlier centuries, but only for a practitioner in Physic (Latin *physica*), that is medicine. Its first appearance in print with a scientific connotation was in 1840 in the preface to William Whewell's book *Philosophy of the Inductive Sciences*. The classification of disciplines was a major interest in the early 19th century and Whewell attempted to classify disciplines in terms of associated activity. He wrote:

> 'We may make such words when they are wanted. As we cannot use 'Physician' for a cultivator of Physics, I have called him a 'Physicist''.

However, Faraday disliked the neologism 'physicist' and he certainly objected strenuously to the suggestion that it should be introduced into the English language. He much preferred to be called a 'natural philosopher' and the same was true of Lord Kelvin, who raised objections to 'physicist' as late as 1890. Indeed, even as late as 1914, scientific journals generally chose to refer to 'men of science' or 'scientific men' rather than 'physicists'.

The word 'science' began to take on a new and narrower meaning in the 1830s and 1840s. It was no longer a synonym for all knowledge (Latin *scientia*), but became instead a description of a particular way of seeking to understand the world. This change of meaning was closely connected to the flourishing activities of the British Association for the Advancement of Science. In a very direct way the Association drew the public's attention to science. Its annual meetings were great events which dominated newspapers for weeks at a time. At its third annual meeting it readily accepted a suggestion from Samuel Taylor Coleridge that its members should not be called 'philosophers'. As a direct result Whewell devised the word 'scientist' to designate collectively those who studied nature whereas previously the word had been used with a much more general connotation, being applied to 'those having knowledge' of any sort. Whewell was responsible in 1834 for the first use *in print* of the word 'scientist' in its more restricted sense.

CHAPTER 2

THE EARLY YEARS OF THE SOCIETY (1874–1890)

In the 1860s and 1870s, a time of enthusiastic rivalry between Gladstone and Disraeli, there was an almost complete lack of government support for scientific research. It is perhaps surprising that the person who did most to get the country to think seriously about scientific and technical research was Colonel Alexander Strange of the Indian Army who, although now an almost completely neglected figure, was a remarkable and far-seeing man. He proposed a system for the organisation of science on a national basis which would operate under a Minister of Science taking advice from two councils, one for military science and the other for civil science—a scheme that was, in essence, adopted almost a century later. J A Crowther, later Honorary Secretary of the Institute of Physics, commented:

> 'Though his ideas were fundamentally correct, they were put out of mind by the British world of science, in a kind of Freudian forgetting, because they were contrary to the individualistic conception of scientific activity then dominant'.

The kind of opposition that faced Strange at this time can be summed up by quoting the reaction of Robert Lowe, the Chancellor of the Exchequer, in 1869 to a request for an extremely modest grant of £300 per annum by the Scottish Meteorological Society to allow the Society to continue making observations. He said:

> 'I am in principle opposed to all grants and it is my intention not to entertain any applications of this nature. We are called

upon for economy ... . I hold it as our duty not to spend
public money to do what people can do for themselves'.

Throughout the 19th century there were fundamental advances in
knowledge. Faraday became interested in electromagnetic phenomena
following Oersted's discovery of the magnetic effects of a current
and Ampère's discovery of the action of currents upon one another.
Faraday's 1831 paper on electromagnetic induction, headed 'Evolution
of Electricity from Magnetism', was profound and far-reaching.
Measurement techniques, always important in experimental physics,
were developing fast: Sir Charles Wheatstone (1802–1875), often
regarded as the practical founder of modern telegraphy, is still famous
for the Wheatstone Bridge, developed by him for the measurement of
resistance. Financial independence enabled James Joule (1818–1889)
to conduct experimental investigations in his own laboratory on the
mechanical equivalent of heat. He continued to refine his measurements
over a ten-year period, reporting periodically to the Royal Society. It
was, however, James Clerk Maxwell (1831–1879), a Scot who put
the empirical work on electromagnetism into mathematical form, who
had perhaps the most profound influence in the second part of the
century. His theoretical work predicted the existence of electric waves
propagated through a dielectric medium, and these were in due course
demonstrated by Heinrich Hertz.

It is against this background of activity that the founding of the Physical
Society of London must be considered. The social and intellectual
climate had led Faraday to originate the series of evening Discourses
at the Royal Institution, and the same climate seems to have been such
as to persuade Frederick Guthrie, who had been appointed Professor of
Physics at the Royal College of Science in South Kensington in 1868,
that it was worthwhile to start a society in what was then a relatively
new and quite small field of science. No doubt the spurt in the building
of physics laboratories in universities in the late 1860s and early 1870s
was a contributory factor. There were also the exciting developments
within physics itself which were helping to delineate boundaries distinct
from chemistry, astronomy and mathematics. In 1873 Clerk Maxwell
published his classic book *Electricity and Magnetism* and was busy
establishing the Cavendish Laboratory in Cambridge in which he was
to become the first professor of experimental physics; John Tyndall
had been appointed in 1867 to succeed Michael Faraday at the Royal

Frederick Guthrie, founder of the Physical Society of London and President 1884–1886.

Institution and he too had made important discoveries in physics; men such as George Stokes, Sir William Thomson, James Joule and Charles Wheatstone were becoming increasingly well known for significant breakthroughs.

The first stirrings towards the formation of the Physical Society occurred in the summer of 1871. Through the initiative of Professor Frederick Guthrie summer courses were started for science teachers from all parts of the country. There was great emphasis upon the construction of simple apparatus, which was hardly surprising because science teachers at that time had little apparatus of their own and very little provision for purchasing instruments. It was probably through this contact with teachers from many different areas that led Guthrie and his assistant Barrett to consider the need for a society which could take responsibility for diffusing knowledge on the progress of physical research. Barrett (who became Sir William Barrett FRS) later commented that he and Guthrie had talked repeatedly about this need for a Physical Society similar to the Chemical Society, which had been formed in 1841.

It was agreed that it would be advisable to obtain the opinions of Fellows of the Royal Society and other well-known scientists as to the support likely to be given to such a society. An opportunity to canvass opinion arose conveniently in 1873 since Barrett was due to give a paper at a meeting in Bradford of the British Association for the Advancement of Science. Barrett duly sent Guthrie a list of those who seemed to be in favour of the idea. It is interesting that of the thirteen Fellows of the Royal Society listed by Barrett only two could be regarded as primarily physicists. The others were chemists, astronomers, mathematicians and a single biologist. As a result of this information Guthrie decided to send a circular from his home address, rather than from his college, to about thirty of his scientific friends and to those who, from their known interests or official positions, he thought likely to be interested in the formation of a society for the cultivation of physical science†. His circular letter began:

> 'I wish to form a Society for Physical Research: for showing new physical facts and new means for showing old ones: for making better known new home and foreign discoveries,

† There was alarm in the Society of Telegraph Engineers when Guthrie sent the letter to fellow scientists proposing the 'new society' to provide a means to exhibit and discuss physical experiments. They feared that members of their society would be lost to this new society. In 1874 a motion was put to the Council of the Telegraph Engineers that in order 'to prevent their taking away our best members, an amalgamation of the two societies be formed'. The Council established a sub-committee to explore the matter, and as a consequence nothing further was ever heard of it.

and for better knowledge, one of another, of those given to physical work'.

It is interesting that Guthrie referred to a 'Society for Physical Research' rather than a 'Society for Physics'. That this was a quite deliberate manoeuvre to increase the likelihood of his suggestion proving acceptable is shown by an early draft of the circular in which he wrote:

> 'It is proposed to form a Society for the purpose of reading and discussing original communications in physics and giving digested summaries of the more important foreign physical researches.   Your presence is therefore requested at a preliminary meeting which will be held at _ _ _ to receive suggestions and organise the proposed Society. _ _ _ has kindly consented to preside'.

It would appear that Guthrie had originally thought in terms of a society for physics as a specified field within science, but that during the preliminary discussions with colleagues he became convinced that it would be advantageous to change to 'physical research'. This term would be taken by most people to refer to methods of investigation. It is also interesting to note Guthrie's deliberate choice of words such as 'showing new facts' and 'new means for showing old ones'. The emphasis was upon 'showing' rather than 'reading and discussing'. From the beginning of the Society there was an emphasis on the experimental side of physical science.   The change from 'physics' to 'physical research' is also evidence that the meaning of the word 'physics' had not yet been universally accepted in Britain. In the early bylaws of the Society it was necessary to include wording that would help to establish the meaning of the term.

Guthrie's proposal met with sufficient support to make it worthwhile to organise a preliminary meeting, even though some people in high places did raise warning voices. For example, in December 1873 Clerk Maxwell wrote to Professor W G Adams of King's College London:

> 'I got Prof Guthrie's circular some time ago. I do not approve of the plan of a Physical Society considered as an instrument for the improvement of natural knowledge. If it is to publish

papers on physical subjects which would not find their place in the transactions of existing societies or in scientific journals, I think that its progress towards dissolution will be very rapid. But if there is sufficient liveliness and leisure among people interested in experiments to maintain a series of stated meetings to show experiments and to talk about them, as some of the Ray Club do, then I wish them all joy; only the manners and customs of London, and the distances at which people live from any convenient centre, are very much against the vitality of such sociability'.

In a continuation of his letter Clerk Maxwell suggests that a feeling of cohesion between members of the proposed society might be encouraged by a regular dinner, possibly with a periodic soirée at which experiments might be demonstrated and described as a basis for discussion. Almost 46 years were to elapse before that suggestion came to fruition. In early 1919 the Physical Society Club was formed at the suggestion of Professor W H Eccles, at that time Senior Secretary of the Physical Society, and it held its first meeting and dinner in London on 23 May 1919. The first minutes of the Club stated that 'a few members of the Physical Society who considered themselves not only sufficiently eminent, or simply erudite, but also congenial, proposed each other to form a nucleus for a club'. Among the original 24 members of the Club were C V Boys, W H Bragg, C C Paterson, E Rutherford and F E Smith. The dinners have been held more-or-less monthly with a gap in the summer months since then and the series will reach its 545th meeting during 1999.

Clerk Maxwell was not the only person expressing opposition to Guthrie's suggestion. There were those who felt that the Physical Society would introduce a divisive element amongst scientists and might damage the Royal Society itself. However, in answer to this point, Guthrie pointed out that fragmentation had already taken place, because by the 1870s there were several specialist scientific societies, including the Chemical Society, which had been founded over thirty years previously.

The first publicity relating to the proposed Society was in December 1873 when both *Chemical News* and *Nature* carried a notice that:

'A preliminary meeting was held on 29 November in
the Physical Laboratory of the Science Schools, South
Kensington to consider the formation of a Physical Society'.

Thirty-six men, including most of the physicists in London, were
present. Barrett, with the full agreement of Guthrie, invited Dr J H
Gladstone FRS to act as Chairman at the meeting and he willingly
agreed. This was a sensible choice as Gladstone was a Fellow of
the Royal Society with a distinguished record in original research
in physical chemistry, especially on the relationship of chemical
composition and optical properties. Thirteen men (no superstition here)
were invited to form an organising committee. A point of particular
importance was that a government organisation, the Lords of the
Committee of Council on Education, granted to the Society the free use
of the Physical Laboratory in South Kensington and its apparatus. This
permission was financially very valuable and it was also of considerable
importance politically because Guthrie wished a characteristic feature of
the Society's meetings to be experimental demonstrations which would
stress that the Physical Society would indeed be different from the
Royal Society.

The Physical Society of London was officially formed on 14 February
1874 at a meeting with 29 people present and at which bylaws
were adopted and the organising committee was declared to be the
first Council of the Society. It was reported that 99 people had
expressed an interest in joining the Society. At first Guthrie deliberately
avoided taking any formal leadership position, but agreed to accept the
modest role of 'Demonstrator of the Society'. The first President was
J H Gladstone, with W G Adams of King's College and G Carey Foster
of University College as Vice-Presidents.

In view of the ambiguous position of physics as a precisely defined
discipline at that time, it was deemed necessary in the bylaws to include
wording that would help to establish the meaning of the term 'physics'.
Under the section 'Name and Objects' the objects were defined as:

'The objects of the Society shall be to promote the
Advancement and Diffusion of a knowledge of Physics. It
is not intended under the general denomination of Physics to
include the details of Chemistry, Astronomy, or the special

First page of the Minutes of the Physical Society of London's first meeting held on 14 February 1874.

branches of Natural Science for the pursuit of which other Societies are already formed'.

It would seem that the framers of these bylaws in 1874 felt that it was not safe to assume that scientists in general would appreciate exactly what 'physics' stood for—in the way that terms such as 'chemistry' and 'astronomy' were tacitly understood. The wording, however, may have been seen as a guarded, negative assertion and in consequence the last sentence was deleted from the objects only two years later at the Annual General Meeting. This may be taken as evidence that the foundation of the Society and its early success had already secured wider recognition of the term 'physics' as a separate field. The range of this field can be realised from its activities and the coverage of the material which was included in the *Proceedings of the Physical Society of London*, which became the journal of the Society.

The early years of the Society (1874–1890)

Technology was beginning to influence the lives of ordinary people in the last decades of the 19th century. London's first telephone exchange was opened. Joseph Swan displayed an electric light bulb with a carbon filament and in 1881 the House of Commons was lit by these bulbs. Thomas Edison meanwhile was working upon an electric light bulb using metal filaments which gave longer life and did not require such a good vacuum. Travel was becoming more sophisticated and convenient with new underground routes through London. This did not prevent some members of the Physical Society from claiming that South Kensington was too far out of town for meetings.

In the years up to 1889 meetings of the Society were held fortnightly at three o'clock on Saturday afternoons from the beginning of November to the end of June in the physics lecture room of the Royal College of Science in South Kensington. Members were allowed free use of the laboratory apparatus in any communication that they wished to make to the Society. These contributions could range from papers describing some piece of research to suggestions for demonstrating well-known phenomena for use in lectures or class instruction. The early contributions were rarely (if ever) subject to any refereeing before presentation. A considerable proportion of the communications were purely oral and were not intended for publication. Special stress was laid on demonstrations in areas of science which happened to be attracting attention at the time. Many small devices for use in teaching were shown.

It is not surprising, therefore, that the communications to the Society during this early period were not always of a high level and indeed sometimes were blemished by quite serious errors. It was, however, suggested that, in those early years, 'the demolition of authors added to the interest and liveliness of the discussions, and no one was much the worse'. Despite this, the Council of the Society was always very diligent in ensuring that nothing of doubtful scientific validity was ever published in its name. It was customary at that time for many members to contribute papers to the *Philosophical Magazine*, which was an independent scientific publication. A good practical arrangement was that those communications which were judged worthy were published in the *Philosophical Magazine* with extra copies of these being printed and then set aside for later distribution. After a sufficient number of these had accumulated they were issued as a volume of *Proceedings*, which also included appropriate administrative details, annual reports, lists of members admitted, and brief obituaries. In due course the *Proceedings of the Physical Society of London* became a regular publication.

In a talk some years later Sir J J Thomson commented that in the 1870s there had been probably fewer than a hundred professional physicists in England. At first sight, therefore, the fact that the Society attracted 99 members in its first year appears remarkable. Close examination of the membership in the first few years, however, reveals that it was very largely made up of teachers in schools and technical colleges and 'amateurs' who had an interest in physics. In addition there were quite a few professors of various related disciplines such as mathematics, astronomy, chemistry and engineering. Indeed, the first President of the Society, Dr J H Gladstone, was at that time Fullerian Professor of Chemistry at the Royal Institution. The membership in the early years was broadly based and not narrowly focused on a restricted range of scientific activities. Furthermore, in taking a broad approach to its activities and in putting emphasis upon the experimental side of physics rather than the mathematical, the Society was being true to Guthrie's original intention for it:

'to take cognisance of smaller matters, points of technical detail, useful laboratory contrivances, experimental methods of illustrating physical principles, questions connected with methods of teaching, and other things of much import'.

(Guthrie's obituary in *Proc. Phys. Soc.* **8** 1886–7.)

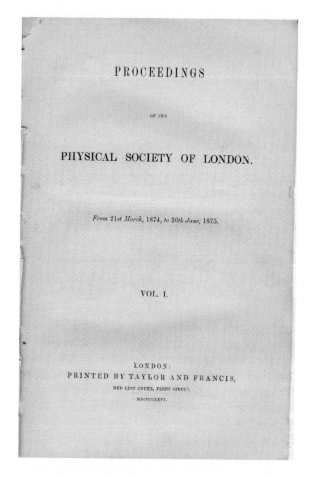

'Proceedings of the Physical Society of London', volume 1.

It is clear that in these early years of its existence the Physical Society made no real attempt to establish a learned society along the lines followed in astronomy, chemistry and geology. It did, however, play a social role. Aside from its regular meetings it engaged in visits such as one to the Cavendish Laboratory in Cambridge and it arranged public meetings for reports on particularly dramatic developments in science and related technical areas. For the members of the Society, there were various publications including a manual on the cgs system of units, an edition of the collected papers of Sir Charles Wheatstone and subsequently an edition of those of James Joule.

Gradually the Society began to attract people prominent in physics. Rayleigh joined within three years of its foundation and Joule within six. William Thomson (later Lord Kelvin) was President from 1880 to 1882 and others amongst the up-and-coming younger generation of physicists were Oliver Lodge, Osborne Reynolds, C V Boys, R T Glazebrook, J H Poynting, Silvanus P Thompson, G J Stoney and G F Fitzgerald, all of whom became early members. At the Annual General Meeting of the Society in 1883 the President, Professor W G Adams, commented on the development of the Society over the decade since its foundation and claimed:

> 'Its history has been that of physical science generally in the United Kingdom for the past ten years, for almost every physicist of eminence has either joined our ranks or has communicated a paper to our meetings'.

It is certainly true that by this time the membership had become representative of the whole community of British physicists: young and old, eminent and otherwise, from a wide range of sub-disciplines of physics, from London and from the provinces.

There was never any doubt that the centre of activity of the Society should be in London and the first three Presidents, J H Gladstone, G Carey Foster and W G Adams, were all London-based; however, the widespread development of the railways after 1850 and a cheap and efficient postal service allowed physicists from all over Britain to be involved. William Thomson (Lord Kelvin) from Glasgow, R B Clifton from Oxford and Balfour Stewart from Manchester were all Presidents of the Society in the first 15 years of its existence. Whilst the interest of distinguished physicists from outside London was important, the attendance figures at the regular meetings of the Society depended upon those who lived fairly close to London. From an analysis of those who attended meetings and the authors of papers in the *Proceedings*, it is clear that there was a relatively small core of members actively involved in the Society. This feature was not uncommon in other associations, which in consequence sometimes became defunct. However, the Physical Society survived and this is perhaps surprising for several

William Thomson, Lord Kelvin (President 1880–1882) (right) photographed with Lord Rayleigh (President 1934–1936) at Rayleigh's home (from 'The Life of Lord Rayleigh' by his son Robert John Strutt, Fourth Baron Rayleigh, © 1924 Edward Arnold).

reasons: first, the regular meetings were held in South Kensington and in those days journeys there were not as easy as today; secondly, the Society seemed completely uninterested in professional matters such as the awarding of designatory letters and the improvement of conditions of work for physicists; thirdly, in these early years there was no advantage in securing publication for a member's paper in the *Proceedings* because if it was of sufficient merit it would already have appeared in the *Philosophical Magazine* which at this time was already a significant journal.

The apparent disdain for professional matters in the Physical Society was in contrast to what was happening with the chemists. The Chemical Society had been established in 1841 and by the 1870s many chemists were urging the need to establish a professional association within the Chemical Society. Pressure for the certification of professional qualifications led in 1877 to the establishment of the Institute of Chemistry. The chemists were active in persuading the government to provide appropriate facilities, to establish awards for commendable chemical projects, and attempting to cajole industry into employing more chemists. Physicists within their own society did none of these things. Unfortunately, in the 1870s and 1880s almost the only future for a professional physicist was in some branch of the teaching profession. In particular, if physicists wished to pursue some physical research they had either to hold a teaching post or to have independent means.

A further difference in the approach to professionalism between the Physical Society and the Chemical Society can be seen in their methods of deciding who should be members. People applying for membership of the Chemical Society were sometimes turned down on the grounds that their qualifications were not suitable. This kind of selective process never seemed to occur in the Physical Society, where election was long and tedious but ultimately a sure process. Potential members simply had to secure the backing of three existing members. The name of the candidate was then submitted to two successive meetings of Council, where objections could be raised. If the candidate survived this test, his name was submitted to a general meeting of members with actual voting being deferred to the next general meeting. Membership of the Society covered many different aspects of science and technology and it was not difficult for a potential member to obtain the backing of three people.

Naturally there were advantages to the Society in these arrangements for election as it brought into membership people from a diversity of scientific and technological interests. Yet there was just enough formality about the procedures to allow the Society to say that its members were in some sense 'suitably qualified'.

It is clear that an early decision was taken and maintained to run the Society as a sort of club for persons interested in physical science rather than as a specialised society for professional physicists. For example, the Third Marquis of Salisbury (who later became Prime Minister) and Thomas Huxley, both of whom had an interest in science, though without qualifications in physics, were nonetheless elected to the Society in the first few years of its existence.

Membership was open to women from the very beginning. In the first decade two women were elected as members and there is no evidence that this was regarded as being unusual. In this respect the Physical Society was in advance of its sister society in chemistry. It was not until 1920 that the Chemical Society submitted a petition to the King for a Supplemental Charter: 'The membership has hitherto been restricted to males, but it is now desired, and is considered expedient, for the welfare of the Society that both males and females should be eligible for membership'.

A change in attitude began after about 1885 when Guthrie, now President of the Society, reported:

'The Council trust that it may become more and more the custom of our Members to communicate their physical work in the first instance to the Society, so that our *Proceedings* may become a fairly complete record of the original work done in England in the departments of physics'.

This suggests a move towards being a learned society, rather than just a means to demonstrate and discuss physical experiments, as was the original object.

Guthrie did not accept the position of President of the Society until 1884 and two years later he contracted cancer of the throat and died on 21 October 1886. This might perhaps be regarded as marking the end of the first stage of the Society's development, for if ever a society owed its existence to the activities of one man it was the Physical Society of London. He is commemorated today by the Guthrie Room in the present headquarters of the Institute of Physics in Portland Place and by the annual award of the Guthrie Medal and Prize.

CHAPTER 3

THE BLOSSOMING OF THE SOCIETY AND THE ORIGIN OF THE INSTITUTE (1890–1920)

The 1890s have been described as the 'high noon of classical physics'. It was also the time when experimental results were to undermine the foundations of classical physics. Nikola Tesla invented an alternating induction motor using a 'Tesla' coil. Heinrich Hertz first produced the radio waves which had been predicted in Clerk Maxwell's theoretical work on electromagnetic waves. Oliver Lodge demonstrated that such waves could be used for signalling and Marconi came to live in England and started calling Hertzian waves 'radio telegraphy waves'. In 1892 Silvanus Thompson was writing about electrical power being transmitted over large distances and Hollerith introduced an electrically driven punched-card system.

But 1895 was the year that Röntgen observed x-rays and Becquerel discovered radioactivity. 1897 was the year that J J Thomson discovered the electron. The Curies discovered polonium and radium, and in 1898 Rutherford showed that uranium emitted two types of particle.

For 20 years after its foundation the meetings of the Society were held in the South Kensington Science Schools, where the authorities had generously provided free accommodation and use of equipment. In 1894, however, it was decided that it would be more appropriate if the Society could secure a footing in Burlington House, the headquarters of

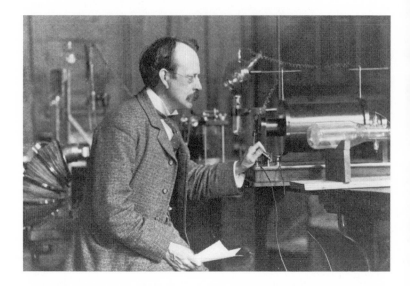

Discovery of the electron, 1897: J J Thomson, first Honorary Fellow of the Institute and President of the Society 1914–1916, with his cathode ray tube (Science Museum/Science & Society Picture Library).

various scientific associations. The Chemical Society offered a cordial welcome and on 26 October 1894 the Society held its first meeting there. It was felt that the new venue was more accessible than South Kensington and closer to other learned societies.

By this time the general affairs of the Society had been put on a more business-like basis. For example, much greater care was taken over the refereeing of papers. No paper was allowed to be read unless it had been referred to and accepted by a competent authority. A practice of putting the more important papers into print before they were read and distributing proofs amongst Fellows known to be specially interested in the topics concerned had also been introduced. These measures led to a distinct improvement in the standard of papers submitted to the Society.

For its first 25 years the Physical Society did not follow the pattern of other scientific societies in requiring the President to give an address at the Annual General Meeting. The first President to fulfil this task was Shelford Bidwell in 1898 and his address included a brief survey of the history of the Society during the previous 24 years. He noted, for example, that the Society's arrangements for meetings had been very resistant to change: the hour and the day of meetings had remained unchanged at three o'clock on Saturday afternoons for 15 years until 1889. Then five o'clock on Friday afternoons proved to be more attractive to younger members, who seemed somewhat reluctant to sacrifice their Saturday afternoons to the pursuit of science. He went on to say that the Council of the Society had more than once given serious consideration to the possibility of holding meetings in the evenings. The balance of opinion had been that afternoons remained preferable and this view had been reinforced by the experience of the Royal Society, which had found that attendance at meetings had increased substantially when the starting time had been moved from eight o'clock to half-past four.

Some evidence of the Society's efficient husbanding of resources can be found in Bidwell's address. The annual subscription for members had remained constant at its original value of £1 until 1896, when a special meeting was held at which a number of resolutions were passed: these included raising the annual subscription from £1 to two guineas; allowing members to be styled 'Fellows of the Physical Society of London'; appointing their own Foreign Secretary and a Librarian, both of whom were to be *ex officio* Members of Council. Despite low fees and a low composition fee for life membership (initially £10, though life members were invited to contribute voluntarily to the Society's funds), the Society's income was so much greater than its expenditure that it had been able by 1898 to accumulate and invest a capital of almost £3000, a very large sum in those days. Despite this capital, the resolution to increase the annual subscription was passed almost unanimously by members attending the meeting. Very few members resigned as a result of the increase!

Bidwell went on to suggest that British physicists were at a serious disadvantage as there was no easily available way of discovering what was being done in other countries and that there was a need for some kind of periodical digest similar in character to the German *Beiblätter*. It was accepted that arrangements for the publication of such a digest clearly devolved upon the Society. The only objection to the production of monthly abstracts of papers appearing in foreign journals was the

Lord Rayleigh (right) and J J Thomson 'at leisure', 1898 (by permission of the President and Council of the Royal Society).

cost, which was likely to be more than the total annual income of the Society. However, a trial period of one year was financed from the Society's capital, supported by grants from the Royal Society and the British Association for the Advancement of Science. The trial turned out to be highly successful and the Council decided to extend the experiment.

In his optimistic report Bidwell felt that there was one respect in which the old times had been better than the new. There had been a lamentable falling off in the number of demonstrations when papers were presented and these had always greatly added to the attractiveness of meetings of the Society. He accepted that a part of the explanation was that, since 1894, the new meeting place had lacked the facilities that had been enjoyed in South Kensington so that demonstrations involved more trouble and usually additional cost. He felt that Fellows should be reminded that the showing of experiments, demonstrations and apparatus need not necessarily be accompanied by the reading of a formal paper and that the demonstrations and experiments did not even have to be new in the sense that they had never been shown elsewhere.

Bidwell gave a short but charming account of his election to the Physical Society which reveals the nature of the Society in its early years. At the time when he first wanted to become a member he did not know anyone who was in a position to support his candidature. He therefore decided to write to one of the two Honorary Secretaries of the Society who kindly gave him an introduction to the President and the two of them agreed to sign his application form and to find the necessary third member for him. He made a large number of friends within the Society entirely as a result of regular attendance at the meetings over several years, and in due course he was elected President. This illustrates that the Society at this time did operate as a friendly club for persons interested in science rather than as a specialised society for professional physicists.

The founding of the Röntgen Society under the presidency of Professor Silvanus Thompson in 1897 was reported by Bidwell with mixed feelings. He accepted that the new society had a wide field open to it, including physiological and medical matters which were not within the domain of physics, but nevertheless he hoped that any researcher making a notable discovery on the physical nature and properties of x-rays would feel that the Physical Society was the proper place to publicise his findings. It is interesting to note that Silvanus Thompson became President of the Physical Society from 1901 to 1903 so relations between the two societies appear to have remained friendly†.

The turn of the century

The developments in physics, begun in the 1890s, continued apace. In 1900 Planck suggested the quantum theory. In 1901 Marconi

† The Röntgen Society was one of the societies, together with the Faraday Society and the Optical Society, which the Physical Society invited in 1918 to join with it in negotiations to set up the Institute of Physics. The Faraday Society and the Optical Society accepted the invitation, but the Röntgen Society delayed doing so until 1920.

Rutherford (right) (President of the Institute 1931–1933) and Geiger counting alpha particles at Manchester (from 'Rutherford: Simple Genius' by David Wilson, © 1983 Hodder & Stoughton).

received messages from across the Atlantic and in 1903 there was the first powered flight by the Wright brothers. Rutherford and Soddy identified radioactivity with the transmutation of elements, Fleming patented the first vacuum tube and in 1905 Einstein explained the photoelectric effect. In 1906 Rutherford identified alpha particles as helium nuclei. Weiss proposed a domain theory of ferromagnetism, Geiger and Rutherford developed the Geiger counter, Onnes liquefied helium and Berliot flew across the English Channel. Rutherford was awarded the Nobel Prize for Chemistry (not Physics), but despite that, and helped to some extent by the Physical Society, the words *physics* and *physicist* were beginning to be recognised in the country.

By 1900 the membership of the Society, as measured by the number of Ordinary Fellows, had risen to 451 and so it entered the new century in a healthy state and during the next 20 years the Society moved along in a serene and pleasant way as a learned society which organised meetings at which a small number of papers were read and discussed. The membership of the Society throughout this period was relatively steady and fluctuated between 407 and 454. The reports at the Annual General Meetings showed that between 10 and 14 ordinary scientific meetings had been held each year with an attendance which varied between 25 and 60 people from the membership of over 400.

For just over ten years most meetings took place in the Chemical Society's rooms in Burlington House. But in his presidential address at the AGM in 1903, Dr R T Glazebrook, the Director of the National Physical Laboratory, deplored an obvious weakness in the Society not having a permanent home. Speaking of the Chemical Society he said:

> 'Our hosts are very kind to us, but a hired lodging is not a home. There is no place to keep the Society's belongings and if a Fellow wishes to illustrate his paper by a demonstration all the equipment has to be collected from a distance and transported to Burlington House'.

He felt that corporate life as a Society was entirely lacking and whilst he could think of no obvious way to remedy the drawback in the immediate future it was important that the Society should bear it in mind. Glazebrook went on to comment on the timing of meetings, which then started at 5 pm, and he raised the possibility of having evening meetings. His words illustrate the nature of the Society at this time:

> 'Most of us are busy men, and at 5 pm have done a hard day's work; when we are tired and want our tea and rest, we have to come here to discuss physics. ... Technical Societies could not meet in the daytime, their members are otherwise engaged. I want the Physical Society to appeal to the same class of men and not merely to teachers and lecturers in physics who are resident in London'.

By 1904 only half the of the 14 meetings of the previous session had been held in Burlington House. The Council was concerned about the absence of equipment and for five of the meetings the Society returned to its original meeting rooms in the Royal College of Science in South Kensington. At the same time Council decided to try alternate afternoon and evening meetings. Initially at least, there did not appear to be any marked preference between the two times but the change in venue did have a marked effect in increasing interest in the meetings. A students' class of membership was also established with the hope that it would act as a feeder to the Society. Between 1905 and 1909 most meetings were held at the Royal College of Science with an occasional meeting at University College, the City and Guilds Technical College in Finsbury, and the National Physical Laboratory. From 1907 light refreshments were provided before afternoon and after evening meetings and this service was specifically aimed at allowing Fellows to become better acquainted with one another. The Society, however, moved its meeting place again in 1910 when almost every meeting, whether ordinary, informal or special, was held in the Physics Department of the Imperial College of Science.

There was also concern about the size of the membership and the nature of the Society. In his presidential address, Dr Glazebrook also commented:

'... it can hardly be argued that the growth of the Society has been commensurate with the interest and importance of the subject. Of late years our numbers have been nearly stationary, and a Society which does not grow, stagnates and will die. Have we fulfilled our appointed task, and must we make way for those more technical Societies which flourish so abundantly? I would say no! Rather let us profit by their example, and claim for ourselves a new and wider range of activity. Physics is a far-reaching subject, one which has contact at many points with other sciences, and our range of papers should be correspondingly great. And yet of recent years, at least, the range has narrowed'.

Glazebrook felt that there was the mistaken feeling that the Society did not deal with technical papers. He suggested that many of the papers read before the Institution of Electrical Engineers would probably be better read before the Physical Society and this view, he claimed, was shared by the current President of the IEE.

In his presidential address in 1908, Dr C Chree commented that with the increased number of people directly or indirectly concerned with science it might be expected that the Society's membership would have increased sharply. The number of Ordinary Fellows had increased by about 7% in six years but certainly not as sharply as might have been expected. Part of the reason for this was the increased specialisation in

science which had led to the rise of new and more specialised societies.
Chree described a further reason:

> 'Considering modern tendencies, one class of the community
> from which I think we might reasonably have expected
> substantial support has accorded it but slightly. Ladies are
> freely admitted to our membership, but for some reason they
> have not hitherto shown much eagerness to avail themselves
> of this opportunity'.

He expressed the hope that the tea and light refreshments might do the
trick!

From its earliest years the Society produced good publications. Initially
it was just the *Proceedings* of its meetings.   In the first decades
contributors included Crookes, Carey Foster, Silvanus Thompson,
Glazebrook, Poynting and Boys. In the following decades the standard
was well maintained with papers by Lindemann, Mott, Raman, Rankine,
Rutherford and Chadwick. Many of Sir William Bragg's papers first
appeared in the *Proceedings*. As might be expected, the number of
papers multiplied enormously as the years went by. In 1910 Council
accepted that only papers read before the Society would appear in the
Proceedings. Fellows were free to offer their papers to any scientific
publication.

Another important activity of the Society began in 1905 when
Council decided to set aside an evening in December exclusively for
an exhibition of apparatus at the Imperial College of Science and
Technology. This innovation was very well supported by manufacturers
and about 240 Fellows and visitors attended the meeting. Thus was born
the Annual Exhibition. In 1910 the 5th Annual Exhibition was open
in the afternoon as well as the evening and in addition experimental
demonstrations were presented by Professors C V Boys and S P
Thompson. These changes resulted in a much higher attendance of
about 700 Fellows and visitors, compared with 200+ in the previous
years.   There were over 30 enthusiastic exhibitors, including such
names as Cambridge Scientific Instruments, Gallenkamp, Griffin &
Sons, Adam Hilger Ltd, Marconi's Wireless Telegraph Co., W G Pye
& Co. and Carl Zeiss Ltd. This new format continued until the 9th
Annual Exhibition in December 1913.

William Bragg (President of the Society 1920–1922 and the Institute 1925–1927) with a portrait of Michael Faraday in the background. Oil painting by Charles E S Phillips (Honorary Treasurer of the Institute 1925–1946) (reproduced by kind permission of the Royal Institution).

By 1913 the financial position of the Society had improved sufficiently that Council felt that the Society's field of activity should be increased. It decided, therefore, to fund from time to time reports on certain subjects of general interest. The first subject chosen was radiation and J H Jeans agreed to write the report which appeared in 1914 with the title *Radiation and the Quantum Theory*. In addition, a Committee was appointed at this time to prepare a publication on *Nomenclature and Symbols* and the report appeared in the *Proceedings* for 1914 so that it could be fully discussed before Council took any definite action on its recommendations.

In 1913 there was a further interesting innovation. Council felt that Fellows would appreciate an occasional lecture by an eminent physicist and the first such lecture was given by Professor R W Wood of Johns Hopkins University, Baltimore, on 27 February 1914 on the topic 'Radiation of Gas Molecules Excited by Light'. This series of lectures was to be known as 'The Guthrie Lecture'.

In 1916 the question of designatory letters for Fellows of the Society was raised. Council decided that since the practice of Fellows had varied from time to time in the past it was highly desirable that there should be uniformity. As a result Council decided to sanction and adopt the letters FPSL as the official indication of Fellowship of the Physical Society of London. Fellows were asked not to use any other designatory letters.

It is interesting to note that the First World War did not appear to have any substantial effect upon the Society's activities apart from the suspension of the Annual Exhibition. Membership remained quite steady at an average of 436 over the four years and the number of meetings held in each session showed little change.

The origin of the Institute of Physics

In autumn 1917, however, a significant step was taken which eventually was to have far-reaching consequences. Council appointed a committee to consider and report upon the possibility of steps being taken to improve the professional status of physicists. In the period up to the outbreak of the war no-one had thought about physics as a profession. Scientists such as Clerk Maxwell and Rayleigh would never have imagined that physics could ever provide a large number of people with the means of earning their livelihood nor that physics would revolutionise the daily life of society as a result of scientific research done in the laboratory. For Clerk Maxwell any close association

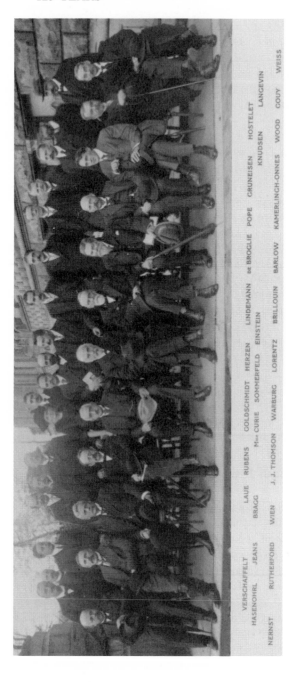

The 1913 Solvay Conference (© Instituts Internationaux de Physique et de Chimie, Bruxelles).

between physics and industrial applications would have seemed as illogical a concept as marrying philosophy and banking. The war, however, had used science to such an extent that the situation was transformed and the idea of a professional organisation for physics became feasible—although there was apprehension on the part of physicists that it would be the chemists and chemistry which obtained the lion's share of status and recognition.

The need for some sort of professional association of physicists first arose during informal discussions in the spring of 1917 in certain committees of the Board of Invention and Research of the Admiralty, where a numbers of physicists were engaged on research work for the war effort. Dr W H Eccles and Mr W B Duddell (the inventor of the electromagnetic oscillograph, but who died later in 1917) started the debate on the possibility of launching an Institute of Physics on similar lines to the Institute of Chemistry. During the summer of 1917 Dr Eccles mentioned the matter to a number of physicists including Professor W H Bragg and Sir Ernest Rutherford, and subsequently he met Captain (later Major) C E S Phillips who suggested that the way forward was through the Council of the Physical Society of London.

There was obviously concern, even dissatisfaction, among physicists at the attitudes towards their profession. Major Phillips told the Physical Society in November 1917 that there was a growing desire, especially among younger physicists, for an improvement in their status. Practical scientific work had been stimulated by the war, and many developments in physics had made important contributions during it, but as the first President of the Institute, Sir Richard Glazebrook, was to say:

'There was little or no recognised position for physicists. Men who have done important work in physics have, in some cases, only been given an official status by being termed Research Chemists'.

These feelings were echoed elsewhere and J H Brinkworth wrote:

'The only heading under which someone with some scientific knowledge could register was that of chemist. As a physicist I deplored this and some time in 1916 I spoke to R S Willows, Secretary of the Physical Society, and to Phillips, one of the younger physicists most concerned, suggesting that for one guinea someone with a degree could be officially registered as a Physicist'.

Phillips felt that the Physical Society was the right body to take the initial steps. He argued that, if it were to be left to the universities, any proposed scheme might suffer too much from narrowing within purely academic limits and if the initial action were to be left to industrial or commercial bodies the scheme might suffer by being out of touch with the educational world. He felt that the best solution was to establish an Institute of Physics with the current Council of the Physical Society forming the first Council of the Institute and the Society's President acting as the first President of the Institute. He proposed that once the ball had been set rolling a new Council of the Institute would be elected and it would then elect its own President and the Institute would become an independent organisation whose main function would be to safeguard the interests and professional status of physicists.

The Council of the Society was persuaded by these arguments and in due course invitations were issued to the Faraday, Optical and Röntgen Societies to join the Physical Society for a discussion on the proposal. A meeting was held in Imperial College on 27 March 1918 to discuss 'the desirability of taking steps to improve the professional status of the physicist by founding an Institution (sic) of Physics'. Present at the meeting were representatives from the Physical Society and three 'participating societies', namely: R W Cooper, Professor W H Eccles and Professor C H Lees from the Physical Society; Professor A W Porter, E H Rayner and F S Spiers from the Faraday Society; S D Chalmers, Dr R S Clay and L C Martin from the Optical Society; and Dr G B Batten, W F Higgins and G Pearce from the Röntgen Society.

At the meeting Professor C H Lees was elected Chairman. The discussion was wide-ranging and included: the possible provision of a headquarters with meeting rooms for use by the participating societies; the formation of a library; the advantages of increasing prestige overseas; exhibitions and publications. Arrangements for the giving of diplomas, the need for registering the qualifications of members, the importance of holding examinations, the value of an employment register, and the setting up of a benevolent fund were all discussed and suggested within the framework of this infant Institute.

In due course a scheme for the establishment of the Institute of Physics proved acceptable to the Physical Society of London, the Faraday Society and the Optical Society and a provisional Board of the Institute of Physics was nominated by those Societies with membership being allocated as follows: seven delegates from the Physical Society and four each from the Faraday Society and the Optical Society. In this way the Physical Society's dominating position in the new Institute was recognised and accepted.

The first meeting of the Board was held on 17 January 1919 at King's College London. Sir Richard Glazebrook was invited to be the first President, Professor A W Porter the first Honorary Secretary and Sir Robert Hadfield became the first Honorary Treasurer. At the May meeting of the Board, a proposal to apply for a Royal Charter was discussed, but was postponed *sine die*, very much so in fact since it was to be another 50 years before the Charter was granted.

The Institute was incorporated under special licence of the Board of Trade on 1 November 1920 but before this incorporation the Royal Microscopical Society and the Röntgen Society became associated with the Institute as Participating Societies and a special clause was added to the Articles of Association to enable them to have representation on the Board of the Institute. The dominance of the Physical Society amongst the Participating Societies was confirmed by the allocation of the members of the second Board for 1921 when the Physical Society nominated two delegates whilst the Faraday Society, the Optical Society, the Royal Microscopical Society and the Röntgen Society each nominated only one. This allocation of delegates continued to the end of 1929.

It is fair to say that the Physical Society, having been very involved in the birth of the Institute of Physics and taking an official interest in its activities through its membership on the Board, nevertheless continued on its own way with nothing radically changed from the programme of events which it had developed during the first two decades of the century. It remained essentially a 'learned society' in everything that it did, being content to leave the professional aspects of the physics community to the Institute of Physics. The emphasis upon activities such as arranging lectures by distinguished physicists, organising discussion meetings on selected topics of current interest, producing a programme of scientific meetings at which members could read papers, establishing an annual exhibition of apparatus, and encouraging prominent authors to produce reports on different aspects of physics all seem to have been attractive to physicists because the membership of the Society as measured by the number of Fellows rose from just under 500 to 1091 during the next 25 years.

At the first Board meeting of the Institute of Physics an oddly named *Propaganda Committee* was appointed, no doubt with overtones of wartime phraseology. Its terms of reference were 'to consider and report upon the means to be adopted to make the Institute generally known'. No doubt the efforts of the Propaganda Committee were successful, for in 1920 *The Times* became aware of the new Institute and its engineering supplement on 20 May carried the following somewhat condescending note:

'The Institute of Physics is to be congratulated upon having managed its inauguration with the least possible noise and friction. By this procedure it has avoided criticism, and it has been able to establish itself without wasting energy upon justifying its existence.

There are obvious reasons for exercising the same judicious care during the next phase of its development, for it occupies a position which will need all the strengthening it can derive from within itself. In the realm of theory, the activities of the new organisation will necessarily overlap those of the Physical Society, but by mutual arrangement there need be but little duplication. Success will depend upon the maintenance of cordial relationship between all the societies and institutions concerned, so that each takes over the work with which it is best fitted to deal.

In promptness of printing and handling papers, the Physical Society in recent years has led the way, and the Institute of Physics will find the Society hard to beat. In the domain of practical physics the difficulties are of a different character. Broadly speaking, applied physics is the province of the engineer, and the Institute of Physics will find itself, alike in professional matters and in the selection and discussion of papers, in danger of being redundant. The antidote must be found in the industrial physical laboratories where the problems of manufacture are entrusted to trained physicists.

To include these physicists may necessitate modification of the rules with regard to those employed in an advisory or consultative capacity, but it is most desirable that provision should be made for accepting them as members, and for ensuring their full representation upon the Board'.

1920 was a busy year of consolidation for the Institute. The Memorandum and Articles of Association were submitted and approved. These broadly defined the scope of the Institute's activities as 'The elevation of the profession of physics and the advancement and diffusion of a knowledge of physics, pure and applied'.

Sir Joseph (J J) Thomson was elected the first Honorary Fellow; the Braggs, father and son, and Charles Parsons were made Fellows. Another election which was to play a large part in future developments was that of Major K Edgcumbe: he was to become the future landlord of the Institute in Belgrave Square.

There was still a fight for recognition. At an April meeting in 1920 it was reported that of 201 people invited to apply for Fellowship, 99 had accepted, 29 refused and 73 did not reply. Attention was being given to overseas matters even at this early date. Professor MacLennan, who was leaving for Canada, was asked to float ideas for a local section there. In July 1920 Dr E H Rayner presented a report on a proposed *Journal of Scientific Instruments*, which was to be discussed by a sub-committee. In October 1920 the Röntgen Society ceased to waver and became affiliated, as did the Royal Microscopical Society.

All this activity prompted further notice from *The Times* and on 23 November 1920 the following more positive report appeared under the heading 'Progress of the new Institute':

'The Institute of Physics has now been incorporated and has begun to carry out its work. Its object is to secure the recognition of the professional status of the physicist, and to co-ordinate the work of all the societies interested in physical science or its applications.

This co-ordination has already been secured by the participation of five of these societies—the Physical Society of London, the Optical Society, the Faraday Society, the Royal Microscopical Society and the Röntgen Society. The first list of members includes the names of over 200 Fellows. Sir J J Thomson, OM, the retiring President of the Royal Society, has accepted the invitation of the Board to become the first, and at present the only, Honorary Fellow.

It is a tribute to the status already acquired by the newly-formed Institute that its diploma is now being required from applicants for government and other important positions requiring a knowledge of physics, and the physicist is now becoming recognised as a member of a specific profession. The first President of the Institute is Sir Richard Glazebrook, FRS, who will preside at the first statutory meeting early in the new year. Particulars with regard to the qualifications required for the different grades of membership can be obtained from the Secretary, 10 Essex Street, London WC2. Fellows elected before May 1, 1921, will have the privilege of being styled Founder Fellows'.

The style of this article probably came from the Propaganda Committee, who reported to a December meeting that a further communiqué had been sent to the press.

By December 1920 the membership consisted of one Honorary Fellow, 206 Fellows, 42 Associates and 57 Ordinary Members. Ordinary members were initially members of the participating societies, but in effect the grade was now open to others.

The inaugural meeting of the Institute

The inaugural meeting of the Institute took place on 27 April 1921 with the President, Sir Richard Glazebrook, in the Chair. Sir Joseph Thomson (who was to become the second President) gave the address. He opened by saying:

'I should like, on behalf of those interested in physics, to express our obligation to those who have conceived the idea

Richard Glazebrook, President of the Physical Society 1903–1905 and first President of the Institute of Physics 1920–1921.

of this Institute, and who have borne the labours in connection with its initiation. The Institute has become necessary, because there is now not an inconsiderable number of men and women earning a livelihood as physicists in one capacity or another. How many people there are in the country who are eligible to be members of this Institute is, I think, a question that is very difficult to decide. We have one fact before us, that in the first year of its existence 300 members have joined. I should think, if I were to make a rough guess and if I were to include among the potential members of this Institute those science masters who have done or are interested in research, there would be around 800 to 1000 available. At any rate, the number is clearly sufficient to justify the physicists in claiming to be a definite and independent profession. And this Institute is one which, like similar organisations of doctors, lawyers, engineers and chemists, has been founded to promote the interest of the profession, to act as a bond of union, to insure that the highest standard of efficiency is reached by those interested in it, and also to insure a high standard of professional conduct'.

He could not resist adding:

'We have heard a great deal about the Institute of Chemistry. Well, here is one who regards chemistry as a branch of physics. It is rather anomalous, to say the least, that there should have been an Institute of Chemistry and not of Physics'.

Further reflections from J J Thomson

He went on to describe the great progress that physics had made and these relections make interesting reading.

'It happens that in a few months I shall have been for fifty years a student of physics, and so the period I have covered is one that has come, so to speak, under my own observation. Well, fifty years ago physical laboratories were very few, and were very sparsely populated. There were laboratories at Oxford, at University College and King's College London, at the Royal School of Mines, at the Royal Institution and the Scottish universities. Fifty years ago the Cavendish Laboratory had been decided upon, but the building had not yet commenced, and I think perhaps that one of the most striking instances I can give of a development of the scale of expenditure connected with physics is to say that the estimate for the Cavendish Laboratory, with equipment, when it was published fifty years ago, was £6,300. Needless to say that estimate was exceeded, but the fact that a very strong and able committee reported that amount to be sufficient is, I think, striking evidence of the difference in scale between fifty years ago and now.

I don't think there were a dozen laboratories—physical laboratories—in the country at that time, and many of these laboratories were exceedingly small. Those I had experience of consisted of a very few rooms snatched by an enthusiastic professor from an apathetic governing body, rooms nobody else wanted and which were handed over to his importuning.

There were few advanced students, and fewer still who intended to make physics the business of their life; and indeed that was a very reckless and dangerous thing because the only positions open to physicists in those days were a few—very few—badly paid professorships, and a few still worse paid masterships at public schools. So that the only people then to make physics a profession were those enthusiasts to whom the delight of research far more compensated for the smallness of their salaries. It was then that physics was its own reward. Physics then was an army with very great generals, but with very few troops'.

He went on to say how things were now changing: 'committee meetings were once sporadic rather than chronic and one had more time for research'. He commented that where a laboratory once had a single induction coil, 12 induction coils were now needed and there were 20 people wanting to use them! But he finished by predicting the ways in which there would be ever-increasing roles for physicists in industry in the years ahead, and the important place the Institute would have in the developments that lay ahead.

Among those at that meeting that day was a fascinated schoolboy named H R Lang, who was to have a prominent role later in the story of the Institute.

CHAPTER 4

THE SOCIETY AND THE INSTITUTE WORKING IN PARALLEL (1921–1939)

Progress in physics continued. Rutherford had proposed a nuclear model of the atom, Bohr had developed his model with planetary electrons and the Braggs had used crystals to measure the wavelengths of x-rays. In 1913 Millikan measured the electronic charge and Moseley developed the idea of atomic number. In 1914 Frank and Hertz confirmed Bohr's model experimentally, in 1917 Einstein produced the general theory of relativity and in 1918 Aston built the first mass spectrograph.

In the 1920s, the Compton effect was discovered, de Broglie introduced the idea of wave–particle duality, Bose–Einstein statistics were discussed and Appleton used radio waves to detect the ionosphere. In 1927 Heisenberg suggested the uncertainty principle and soon afterwards Dirac predicted the existence of antimatter and Pauli postulated the existence of the neutrino. In the 1930s, C D Anderson discovered the positron and Chadwick the neutron, and Lawrence produced the first cyclotron. In 1932 there had been the first transmission by the BBC from the new Broadcasting House and a newspaper reported that John Cockcroft and Ernest Walton, under the leadership of Sir Ernest Rutherford, had 'split the atom' at the Cavendish Laboratory. In 1936 Hubble predicted the expansion of the universe and in 1938 fission of uranium was achieved.

Elsewhere, Amy Johnson, a 26-year-old who had gained her pilot's licence only a year earlier, was the first woman to fly solo around the world. The *Highway Code* was drafted and the Youth Hostels

The Cockcroft–Walton accelerator at the Cavendish Laboratory. John Cockcroft (President of the Institute 1954–1956 and first President of the Institute of Physics and the Physical Society 1960–1962) is sitting in the counting chamber (from 'Rutherford: Simple Genius' by David Wilson, © 1983 Hodder & Stoughton).

Association was formed. John Logie Baird demonstrated sending live pictures by television. Going to the 'pictures' was what literally half the population did weekly. Malcolm Campbell set a new land speed record of 301 mph in 1935, while in 1931 traffic lights had been introduced throughout the country to slow everyone else down. James Joyce's novel *Ulysses* was published in a limited edition and promptly banned on grounds of pornography.

The Physical Society

The prestige of the Physical Society was high, not only in the United Kingdom but also abroad, and this may be illustrated by the physicists invited to give the annual Guthrie Lectures. In the 1920s the list reads: Charles Guillaume (Nobel Prize for Physics 1920), Professor A A Michelson (Nobel Prize for Physics 1907), Professor N Bohr (Nobel Prize for Physics 1922), Sir James Jeans, Monsieur le Duc de Broglie (Nobel Prize for Physics 1929), Professor W Wien (Nobel Prize for Physics 1911), Professor C Fabry, Sir Ernest Rutherford (Nobel Prize for Chemistry 1908), Sir Joseph Thomson (Nobel Prize for Physics 1906), Professor P W Bridgman (Nobel Prize for Physics 1946).

The scientific meetings on particular topics became an increasingly important part of the Socierty's annual programme. They occurred up to four times a year and a wide range of topics was covered, such as the teaching of physics in schools, relativity, x-ray spectra, the making of reflecting surfaces, lubrication, metrology and its application to industry, and all these proved popular with attendance figures often over a hundred, approximately twice those for the ordinary meetings at which members' papers on different topics were read. Normally up to 14 of these ordinary meetings were held in each year with anything from 30 to 60 papers and up to 15 demonstrations being given. Attendance at these meetings would usually be about 60. Throughout the period to the beginning of the war in 1939 almost every meeting of whatever kind was held at the Imperial College of Science.

In 1924 the Society repeated a venture which had first taken place in 1914 when it held a meeting outside London. Sir Ernest Rutherford again invited the Society to come to Cambridge. In the morning members visited the Cambridge Instrument Company. After lunch

in Trinity College, the Society held its ordinary scientific meeting
in the Cavendish Laboratory.   In 1925 about 180 Fellows visited
Oxford at the invitation of Professor F A Lindemann.   The Lunch was
in Christ Church and tea in Wadham with the science meeting in the
Clarendon Laboratory.   At least the Society was making an effort to
move outside London, though it was not until 1928 that it moved
to provincial venues other than Oxford or Cambridge.   In that year
members of the Society and their friends paid a one-day visit to
Bristol University at the invitation of Professor Tyndall and in the
following years similar visits were made to Birmingham University,
the British Thomson–Houston Company at Rugby and Rugby School,
Reading University, Nottingham University College, the Cambridge
Instrument Company and the Cavendish Laboratory (for the third
time), the Royal Naval College Greenwich, Royal Holloway College
at Englefield Green in Surrey, the Research Laboratories of the
General Electric Company at Wembley, the Fuel Research Station at
East Greenwich and the National Maritime Museum, Greenwich, the
Cambridge Instrument Company and the Cavendish Laboratory (for the
fourth time), and Southampton University College. These very pleasant
and popular meetings had to cease because of the outbreak of the war
in 1939.

The Annual Exhibition, which had been suspended during the First
World War, was resumed in January 1920, this time a two-day meeting
in cooperation with the Optical Society.   About 40 firms took part
and over 2000 Fellows and visitors attended.   Two special lectures
were given which proved to be very popular.   Between 1923 and
1929, and again in 1931 and 1932, the Exhibition was described
as being put on jointly by the Physical Society and the Optical
Society.   It became increasingly successful.   Whilst the Exhibition
remained a two-day affair, the number of people attending grew to
about 3000 in 1925 with the number of firms involved growing to
60.   From 1926 onwards the Exhibitions became three-day affairs
with more and more firms exhibiting until by 1930 there were over
80.   The increase in the Exhibition's scope was not confined just to
having an extra day on which the general public was admitted: a
new Research and Experimental section was instituted to allow the
results of recent physical research, improvements in laboratory practice
and interesting historical experiments in physics to be shown.   This
section was aimed at research associations, government and industrial

laboratories, university laboratories and private individuals, and it remained popular until the Exhibition had again to be suspended in 1940 due to the Second World War. A further extension was made in 1930 when a new section was started with the object of encouraging craftsmanship in the scientific instrument trade. This section allowed the work of apprentices and learners to be shown in a competitive atmosphere, with awards being made for the best exhibits. It showed that the Society was serious in seeking to extend further its influence beyond the rather narrow confines of purely academic physics in universities.

At each of the Exhibitions there were two or three special lectures or discourses which attracted visitors. Overall, during this period from 1920 to 1939, the number of people attending was very substantial and reached a maximum of about 9000 at the Exhibition of 1935.

Since its birth the Society had been anxious to disseminate knowledge about physics through its publications. The *Proceedings* was, of course, a prime source for this. Early in 1921 Council considered the size of the page, the typeface and the quality of the paper used for the *Proceedings* and decided to make changes so that they would become very similar to the *Proceedings of the Royal Society* and identical to the *Proceedings of the Optical Society*. The *Proceedings* went from strength to strength, and played an important part in boosting membership. In the year 1924–25 the number of pages published in the *Proceedings* was 356 and the total membership was 651, whereas only ten years later in 1935 the corresponding figures were 1154 pages and 962 members. But the efforts of the Society in publication went beyond this. There were the reports on special topics in physics such as *The Quantum Theory* by J H Jeans, *Series in Line Spectra* by Professor A Fowler and *Band Spectra of Diatomic Molecules* by Dr W Jevons.

In 1934 the Society published the first of a series of annual books with the title *Reports on Progress in Physics*. It contained reports on the various main branches of physics together with a number of reports on special subjects. The volume ran to 370 pages and had the same format as the *Proceedings*. The price for copies bound in cloth was 9s 0d to Fellows of the Society and 12s 6d to non-Fellows. This publication proved to be very popular and it was published by the Society every year, even during the war. The viability of these annual reports, and

The Society's Jubilee banquet in the Connaught Rooms, 22 March 1924.

also of the *Proceedings*, was increased substantially by the fact that the American Institute of Physics was extremely cooperative in bringing both of these publications and the advantages of membership of the Society to the attention of American physicists.

A number of important events took place in the inter-war years. During three days in March 1924 the Society celebrated its Jubilee. On 20 March the President, Mr F E Smith FRS, opened the proceedings in the Lecture Theatre of the Institution of Electrical Engineers by welcoming delegates from all the important scientific and engineering bodies in the UK, delegates from the Royal Society of Canada, the Royal Societies of Victoria and South Australia and the Royal Dublin Society, and delegates from scientific societies in Europe, America and Japan.

There were still alive two Founder Fellows and three other Fellows who had been associated with the Society since its earliest times. On 21 March (the actual anniversary of the first ordinary meeting of the Society) the Society heard from three of them. The first was Sir William Barrett FRS, who had been so closely involved with Guthrie in forming the Society. The second was Professor J A Fleming FRS, who had had the distinction of reading the first paper before the Society with the

Proceedings of the Society's Jubilee Celebration Meetings, 20–22 March 1924.

title 'On a New Contact Theory of the Voltaic Cell'†. The third was Sir Arthur Schuster, who had been President of the Society from 1912–1914, who spoke on his reminiscences of famous scientists that he had known. On 22 March a banquet was held at the Connaught Rooms. Amongst those attending were HRH the Duke of York (later to become King George VI), Ramsay MacDonald (the Prime Minister), Viscount

---

† It is interesting to note in passing that, on 28 April 1939, Sir Ambrose Fleming, the Senior Fellow of the Society and the only survivor of the original 99 Fellows, gave a paper and demonstration on a novel method of electrification more than 65 years after reading his first paper!

Haldane (Lord High Chancellor), Sir Richard Glazebrook (Editor of the *Dictionary of Applied Physics* and sometime Director of the NPL and the first President of the Institute of Physics), Professor Charles Fabry (President of the Société Française de Physique), Sir Oliver Lodge (Senior Past-President of the Society), Sir Joseph Thomson (Master of Trinity College, Cambridge), The Rt Hon Sir Joseph Cook (the High Commissioner of Australia), J H Jeans (Secretary of the Royal Society), and Sir Ernest Rutherford (Cavendish Professor of Experimental Physics in Cambridge). Clearly a splendid occasion.

In 1932 the Physical Society of London and the Optical Society were merged. This step was desirable on a number of grounds and the Councils of the two Societies recommended an amalgamation. There was an overwhelming majority when it was put to the vote. The necessary legal business was done and the fusion of the two Societies was completed under the title of 'The Physical Society'. Changing the title of the 'Physical Society of London' had been considered previously in 1926, but Council at that time decided to take no formal action about deleting the words 'of London'. Instead it adopted a procedure, similar to the Royal Society, of not giving prominence to the two words. From 1926 onwards the words 'of London' were omitted from all publications and printed papers, except where legally necessary since the name of the Society 'The Physical Society of London' was given in the Memorandum of Association.

The union of the two Societies meant that the Physical Society had an increased membership and correspondingly increased responsibilities. It was agreed that activities which were specially characteristic of the parent societies would be continued. Thus the Guthrie Lecture and the Thomas Young Oration would be delivered at regular intervals as in the past. On the other hand, the *Proceedings* would in future reflect the amalgamation and would be issued in six parts per annum instead of five and would include optical and allied matters.

An International Conference on Physics organised jointly by the Royal Society and the Physical Society was held in October 1934 in conjunction with a meeting of the International Union of Pure and Applied Physics. This conference was the first of its kind to be held in England. Meetings for the discussion of papers on nuclear physics and the solid state of matter were organised in London and Cambridge.

He made special reference to the services of the
retiring President, Prof. O.W. Richardson, and of
Sir Arthur Schuster, who was retiring from the
office of Foreign Secretary. Dr Vincent seconded,
and the vote was carried unanimously.
Prof Richardson expressed his personal thanks,
and expressed appreciation of the cooperation
of his fellow officers during his period of office.
    Mr T. Smith proposed and Dr Thomas
seconded a vote of thanks to the Governors
of the Imperial College for the facilities
provided for the Society's meetings in the
College. This was carried unanimously.
    A Ballot took place for the election
of Prof. Albert Einstein, who had been
duly nominated by Council, as Honorary
Fellow of the Society. The result of the ballot
was that Prof. Einstein was elected unanimously.
    The Scrutators then reported to the
President, who announced the names of those
elected as Officers and Council as follows:—
            (over page)
    The new President, Dr W.H. Eccles
then took the chair, and expressed thanks
for the honour of election as President.
    The meeting then adjourned.

*Albert Einstein was elected an Honorary Fellow of the Society at the Annual
General Meeting on 23 March 1928.*

About 30 papers contributed by authors from 10 different countries
were read. Six hundred people enrolled as members of the conference
with 200 from overseas.

The Institute of Physics

Publishing was very much in the mind of the Board of the Institute of Physics. It was clear to the Board from the outset that it needed a publication and in May 1922 a preliminary number of the *Journal of Scientific Instruments* was published. Dr J S Anderson from the National Physical Laboratory was its first Editor and regular publication began in October 1923.

By 1924 Sir Charles Parsons had succeeded Sir Joseph Thomson as President of the Institute. He was born in 1854, the sixth and youngest son of William Parsons, the Third Earl of Rosse. A man of immense talents, Parsons was probably most famous for his work in devising the first steam turbine to work effectively. At the Spithead Naval Review of 1897, his revolutionary 100-foot ship, *Turbinia*, moved at a sensational 34.5 knots. He then turned his attention to optics and designed searchlights and large telescopes. Charles Parsons is still remembered on Tyneside, where he served his first apprenticeship and later founded his own works, and *Turbinia* is on permanent exhibition at the Tyne and Wear Museum.

At the Annual General Meeting of May 1925 Major C E S Phillips was elected Honorary Treasurer, an office he held until 1945. The name Major Phillips runs through the early history like a thread. As early as 1917, before the Institute was formed, he spoke eloquently for its need at meetings of the Physical Society. He was among the first Guarantors; he served on the Board, on committees and sub-committees, and he made various visits as a representative of the Institute. Whenever a fresh field needed investigating, his name appeared. He was an assiduous honorary officer of the Institute and one of its creators. His portrait at the Institute, painted by Sir William Nicholson RA and presented by Major Phillips's widow in 1953, shows at first glance a sober and serious appearance with a clipped military-style moustache, an impression at once belied by the slightly jaunty set of a colourful bow tie and a humorous mouth. His sense of fun is evinced in a letter to Professor Alfred Porter, then Honorary Secretary, when speaking of his OBE. He wrote: ' The latest description of the OBE may be new to you, the 'Old Buffers Encouragement'. It is less coarse than 'Order of the Bad Egg''. That was written in July 1919; other interpretations have appeared more recently!

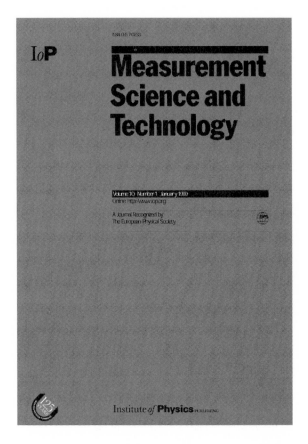

'Measurement Science and Technology'. 'Journal of Scientific Instruments' (first regular issue, October 1923) became 'Journal of Physics E: Scientific Instruments' in 1968 and 'Measurement Science and Technology' in 1990.

When the Institute was founded, it was suggested that it should have a Benevolent Fund and such a fund became an aim of the Board in 1924. The generous Major Phillips donated £100 to start it, and it was suggested that members should pay an annual sum of 2s 6d towards the Fund†.

† The value of the Benevolent Fund has risen to over £800,000 by 1999 and it continues to give help to physicists or their relatives in need.

Charles E S Phillips, Honorary Treasurer of the Institute 1925–1946.

The office work for the Institute was covered at this time by F S Spiers, who was the first Secretary to the Board. He came from the Faraday Society and worked from their office in Essex Street off the Strand. It was proving difficult to set up a separate office for the Institute. Major Phillips inspected various possibilities, but he reported to the Board that the premises inspected were too costly. The problem was repeatedly discussed in Board meetings in 1924 and 1925.

At a special Board meeting in May 1926, the members stood in respect on hearing of the sudden death of F S Spiers. Frederick Spiers had been an important presence since the earliest days, first as one of the representatives from the Faraday Society at the founding meeting in 1918 and later as the first Secretary to the Board. He had served during the formative years of the Institute and was a special loss when large-scale developments were being planned. The membership of the Institute had almost doubled in his time and the consequent administrative work meant that more room space was urgently needed.

1 Lowther Gardens

On 1 May 1851 Queen Victoria had opened the Great Exhibition in Hyde Park. Devised by her beloved Albert, it showed some six million visitors an immense range of scientific, engineering, manufacturing and fine art from Britain and the world. Naysmith's steam hammer, fine Wedgwood pottery and the Koh-i-Nor diamond were all housed under Joseph Paxton's glass Crystal Palace, newly built in the Park; the displays were all intended as demonstrations of British technological and craft supremacy. A Royal Commission had organised it from 1 Lowther Gardens, around the corner from the Albert Hall. It was to these Commissioners that the Board of the Institute turned in 1926 in its search for new accomodation.

Of the participating societies, the Physical Society and the Optical Society wanted to avail themselves of the new accommodation and a joint secretariat; the Faraday Society wanted an office at the new premises for its Editor; the Royal Microscopical Society, which had now become a participating society, wished to carry on in its own premises.

Negotiations were swiftly concluded and by November the offer from the Commissioners to let the Institute have a three-year tenancy from January 1927 of the second floor at 1 Lowther Gardens was unanimously accepted. The terms were extremely favourable: no rent, rates or taxes, payment only for decorating and carpeting the staircase and the rooms. There was obvious delight at the prospect of new offices. Sir William Bragg had succeeded Sir Charles Parsons as President and he presided over this time of change with proper office facilities for the Institute. There was a continuing rise in membership (giving the Institute status and removing the sobriquet 'new'). Thomas Martin was appointed as the new Secretary to the Board and took up his appointment in January 1927. Herbert Lang, the schoolboy who had watched excitedly when J J Thomson had lectured at the inaugural meeting in 1921, was elected an Associate at the Board meeting in February 1927.

So began years of much activity in Lowther Gardens. A joint library was established and a Registrar was appointed to cope with the expanding membership and to revise the regulations for membership. The organisational work for the Annual Exhibition of the Physical Society was based there, as was the editorial work for the Physical Society's publications. Then, in the midst of the bustle, the Board lost its Secretary yet again as Thomas Martin took up an appointment as General Secretary of the Royal Institution. Dr J J Hedges was appointed as the new Secretary in 1927 and work continued.

There was a juggle amongst the Participating Societies. The British Institute of Radiology incorporated with the Röntgen Society and was admitted to the Institute, as was the Royal Meteorological Society. The Royal Microscopical Society withdrew.

The appointment of Dr H R Lang

During 1930 the Honorary Secretary presented a report to the Board on the serious delay which had occurred in the publication of the *Journal of Scientific Instruments*, and intimated that the Officers had found it necessary to direct Dr Hedges to cease acting as Secretary of the Institute and Editor of the Journal. The Board confirmed the action of the Officers in the matter. It was agreed that Dr Hedges be invited to resign his Fellowship of the Institute. The Board considered the position arising from the termination of Dr Hedges' appointment and decided to ask Dr H R Lang to act as Secretary of the Institute until the end of January 1932, at a salary of £8 per week 'so that the arrangements of the Exhibition should not be interrupted'. Dr Lang agreed to this. A sub-committee was then set up to select a new Secretary in succession to Dr Hedges. As might be expected, Major Phillips was part of this selection committee and the outcome was that Dr H R Lang was appointed as the new Secretary. Thus began an appointment which was to last for the next three decades, a vital factor bringing stability to the Institute in the problems that lay ahead.

Herbert R Lang, Secretary of the Institute 1930–1965.

Expansion of the Institute's activities

Some scientists were looking for greater representation in government matters. After a conference on the subject, the British Science Guild, a body described as 'bringing together men of science and men of public affairs for mutual advantage', asked for a representative from the Institute to help set up a Parliamentary Watching Group in association with the Association of Scientific Workers. The ever-watchful Major Phillips advised caution. He pointed out that the Institute needed to preserve a free hand for future dealings in which it might wish to take initiatives outside the Guild, far-seeing advice in view of future growth.

In 1932 Lord Rutherford was President and Professor J A Crowther took over from Professor A O Rankine as Honorary Secretary. Major

Phillips continued as the Honorary Treasurer and Dr H R Lang was now firmly established as the Secretary to the Board of the Institute. The lease of the offices in Lowther Gardens was again renewed on the same terms and Lord Rutherford opened the newly established reading rooms there. The confidence of the Institute was shown by issuing a scale of minimum fees for consultancy work.

The Board noted and greeted the arrival of the American Institute of Physics. Back in 1919, Professor MacLennan, leaving for Canada, was asked to float ideas for a section there, but sadly had to report back that the Canadians had decided against it. Branches were established in Australia and India. With these overseas branches established, attention was turned to branches at home. The first was the Manchester Branch in 1932 with Lawrence Bragg as Chairman and Dr H Lowery as Secretary.

With the arrival of the Manchester Branch and thoughts about a Midland Branch, it became necessary to revise fees for Ordinary Members, at the same time granting them eligibility to join local branches. A quite serious loophole in the Articles of Association became apparent in 1933. It had recently been decided that Fellows should receive copies of the *Journal of Scientific Instruments* without charge. An Associate demanded from the Institute a free supply of the journal, as was available to Fellows. He quoted from the Articles and the Secretary was instructed to make a fairly routine reply, but this did not end the matter. The member reiterated his demand, further correspondence resulted from the Honorary Secretary and the Board's solicitor became involved. It required a Special General Meeting to decide future entitlement of members to publications, and the matter was effectively ended.

The first small steps towards amalgamation of the Institute with the Physical Society some 30 years later were made in 1933. Professor J A Crowther, the Honorary Secretary, spoke of a 're-consideration of the relationship between the Society and the Institute during the coming year' and the arrangements under review with the Physical Society were implemented from January 1935. Obviously the two organisations were becoming more intermingled and the plans included provision of Institute staff to carry out the Society's work including finance, publication sales, advertisement and exhibition organisation.

Laboratory arts including glass-blowing also came under review in 1933. Suitable courses in this country hardly existed, but the following year saw a scheme of training for the Institute's 'Certificate in Laboratory Arts' being set up in Manchester College of Technology.

A Midlands Branch was established with the inaugural meeting at Birmingham University on 1 November 1935. The outstanding success of 1935 must have been in Manchester, where the branch, under the Bragg and Lowery partnership, arranged a conference on industrial physics. The specific subject was 'Vacuum devices in research and industry'. The total attendance over the three days in March was 543. An exhibition of instruments, apparatus and books cognate to the subject was held in the Physics Department of the University and was visited by some 3500 people, figures impressive even by today's standards. A catalogue was published and press coverage was most successful, with notices in daily as well as the technical papers. A detailed report was published in *Nature* on 6 April 1935.

For some months the Board had been represented among the rarefied political atmosphere of parliamentary committees, but in January 1936 it came face to face with *realpolitik*. The British Union of Fascists contacted the Board with a request that members of the Institute should attend a dinner to hear Sir Oswald Mosley speak on 'The structure of Government under British Fascism and its relationship to Scientist and Technician'. The Board, with great aplomb, agreed that no action be taken.

Sir Richard Glazebrook died in December 1935, aged 81:

> 'After spending his early years studying at the Cavendish Laboratory in Cambridge, he was later appointed a demonstrator there and then assistant director. With Sir Napier Shaw he published the *Textbook of Practical Physics* before leaving to become Principal of University College, Liverpool. The newly established National Physical Laboratory chose him as its first director in 1899. On its formal opening in 1902 the NPL comprised only two departments with a staff of twenty six. By the time of his retirement in 1919, the staff numbered over five hundred. Posts at Imperial College and textbooks on heat, light,

mechanics and electricity followed. He was President of the Physical Society in 1903–5; President of the Institution of Electrical Engineers in 1906–7; President of the Optical Society in 1911–12; he was the first President of the Institute of Physics (1920–21)'.

Work at Lowther Gardens continued to expand. The lease was again extended on the same terms and requests were made for extra space at Imperial College. One reason for the increase in work was no doubt because of the variety of activities being organised. Physics was becoming of wider interest in community life, one instance being a lecture organised at the Royal Institution by L C Nickolls of the Hendon Police College on 'Physics and the detection of crime'. Another was the production of a list of members which had a very professional look with general details of members and their posts. There was respect for Associateship as being equivalent to an honours degree for employment purposes. Press coverage was clearly becoming significant.

Facilities arranged by the Institute for students to gain experience in industrial laboratories got off to a somewhat fragile start, with the following pithy reaction from the director of one Research Association:

'... the impression one gains from the past two years is that in university training theory takes precedence over practice, in which respect our own assistants who are working for qualifications by evening study show a decided superiority. This rather theoretical outlook seems to be represented in a lack of attention to detail and neatness'.

The Institute was involved in a deputation to the Postmaster General on the question of suppression of radio interference. Needless to say, the nominee from the Institute of Physics was Major Phillips.

The second of the Institute's industrial physics conferences was on 'Optical Devices in Research and Industry' organised by the Midland Branch and held in Birmingham in March 1937. Although attended by fewer people than the Manchester meeting in 1935 (414 against 543, and 2500 against 3500 at the Manchester exhibition), the conference was held to be an undoubted success.

Lord Rutherford, President of the Institute 1931–1933.

The Institute's bankers were situated then as now in Knightsbridge, but more humbler means of thrift were not neglected as the Board had noted:

> 'That the following members of the Board of the Institute be and are hereby authorised to sign notices of withdrawal of moneys from the account or accounts kept in the name of the Institute with the Post Office Savings Bank'.

On 20 October 1937, Lord Rutherford died. The following is an extract from an obituary:

> 'One of the greatest experimental physicists, Ernest Rutherford was one of twelve children in a farming family

in New Zealand. After winning a scholarship to Canterbury College, Christchurch, he gained an exhibition to work at Cambridge under J J Thomson. His work on radioactivity at McGill University, Montreal, where he was appointed to the Macdonald Chair in 1898, and later when he accepted the Langworthy Chair in the University of Manchester are among the greatest researches of the world. After receiving the Nobel Prize for Chemistry in 1908, he returned to Cambridge in 1919, succeeding Thomson as Cavendish Professor of Physics. He became director of the Royal Society Mond Laboratory in 1936. Many tributes were of course paid to his devotion and contribution to science, but perhaps the most delightful reminiscences were from his sisters when they told of his boyhood bird-nesting, spearing eels, catching trout and earning money on hop-picking holidays, and how, when being asked to teach his sisters during the holidays he would keep their attention by tying their pigtails together. Memories of Rutherford present him as a man of awesome ability and great humanity'.

CHAPTER 5

THE WAR YEARS (1939–1945)

The summer of 1938 was a good one. Families enjoyed their holidays at Margate or Blackpool, students and the young cycled or hiked with the YHA across Europe, and Len Hutton made a record innings of 364 against the Australians at the Oval. But the shadows of war were lengthening and the Institute became very much aware of them when a request came from the Ministry of Labour in October to cooperate in the establishment of a central register of members willing to serve the government and industry in a time of emergency. The response to the circular letter sent to members was reported at the next meeting of the Board with over 300 offers of service and replies still arriving.

Discussions now began in earnest on the possibility of war. Air-raid precautions were taken, and all 'provisional emergency arrangements made to enable the Institute's work to be carried on'. The 'Third Conference on Industrial Physics' at Leeds went ahead, but only after much consideration given to its postponement. It was a slightly muted affair, but it did, however, attract an audience of 232 and 23 firms participated in the related exhibition. An international conference commemorating the discovery of radium was held in Paris late in 1938. This must have caused worry, but eventually the Honorary Secretary, J A Crowther, and the Vice-President, G W C Kaye, were appointed as Institute representatives.

Of course ordinary matters had to continue at the Institute. The death of A W Porter, the first Honorary Secretary, was reported to the Board in March 1939; the Australian Branch chafed at delays in dealing with applications sent to the Membership committee and the Board decided in future to deal directly with them rather than put them to the committee.

A high tension generator being demonstrated at the Radiological Exhibition at Central Hall, London, in December 1938 (Science Museum/Science & Society Picture Library).

However, arrangements for war were hardening. What had become the national register of physicists was being constantly amended by requests from members wishing to register for war work if needed. A committee, including Major Phillips, was set up on how to classify a 'physicist' for inclusion in a schedule of reserved occupations. Although the lease had been renewed at Lowther Gardens, J A Crowther offered office accommodation at the University of Reading and later in 1939 the evacuation went ahead. A National Emergency Committee of the Board was set up to consider how the Institute might best be utilised in the event of the conflict that now seemed inevitable.

Perhaps, however, with Britain and Germany steeling themselves for war, the Board's priorities were best shown in the note after the last

Minute of the Board meeting in May 1939:

> 'In accordance with rule 9 of the Fund (Benevolent) this Committee reports that it has made a grant of £25 to the widow of a distinguished German physicist. The application for assistance was made on her behalf by Sir J J Thomson'.

The early years of the war were difficult ones for the Institute. The inevitable disruption following the relocation to Reading was further complicated by having to recruit new staff locally. All the London staff, except the Secretary (Dr Lang) and one assistant, were either called up for military service or seconded to new posts. Dr Lang himself was deferred from service until 1943 when the Board would re-apply for his deferment to continue.

The Physical Society had moved its ordinary business to 1 Lowther Gardens early in 1939 and when the Institute transferred its office to Reading, it meant that the financial and working arrangements between the Institute and the Society were suspended for the war period. However, a small staff continued to operate at 1 Lowther Gardens throughout the war. The staff was so limited, particularly on the male side, that a successful appeal was made to Fellows for help with fire-watching duties as part of the necessary air-raid precautions!

It proved possible to continue the publication of the *Proceedings* and the *Reports on Progress in Physics* throughout all the years of the war. The Society had been prudent in purchasing on favourable terms a stock of high-grade paper and of binding cloth which lasted for about two years. In the later years of the war, however, difficulties over paper and cloth became acute. The Society was quick to acknowledge how much it owed to the encouragement that it received from American physicists, where subscriptions to both publications and elections to Fellowship steadily increased.

Ordinary science meetings of the Physical Society were suspended for a few months, but restarted from February 1940 at approximately fortnightly intervals though noticeably fewer papers were presented. The concentration of meetings in the Physics Department of Imperial College had now to be varied because of the war and meetings were also held in Bedford College, Birkbeck College and the Royal Institution.

Many features of the normal programme had to be cancelled. The most obvious casualty was the Annual Exhibition of Scientific Instruments and Apparatus and the associated discourses. The 29th Exhibition in January 1939 was the last until the war was over. The first Rutherford Memorial Lecture was postponed, as was a joint meeting with the Science Masters' Association for a discussion on 'The teaching of physics in schools'. The 1940 summer meeting in Oxford was cancelled and so were two series of meetings in northern universities. The Society was having to carry out its work under great difficulties with a reduction of income and it was greatly heartened by an anonymous donation of £500 by a Fellow (a very considerable sum in those days).

In spite of cramped space and awkward conditions in Reading, the Institute became a central clearing house for incoming and outgoing information. Its role in maintaining the physics division of the Central Register for the Ministry of Labour and National Service was crucial. On the other hand, it was still a professional organisation and, while offering every assistance to the Ministry, refused to make confidential files available. This must have been a balancing act for the Secretary. He was in daily touch with the governmental war machine through to 1945, and no doubt under pressure from it, but he was still aware of his responsibility towards the Institute's members. Also, the office was acting as a bureau of information for members wanting guidance as to how their expertise could best be used. Information was sent periodically to members telling them of their situation with respect to the National Service Act and dealing with other professional enquiries. Dr Lang was fortunate at this time to have Board members who were not only completely supportive of him, but who themselves were playing important innovative roles towards the winning of the war.

Everyday problems encroached on the Institute. Censorship, paper rationing and transport problems made for difficulties in publication production. One result was the reduction in the size of the *Journal*. An aspect of the war perhaps not always appreciated was the difficulty in travel for civilians. Restrictions made even simple journeys tiresome, particularly in the early years of the war, though this eased later. Professor J D Cockcroft and R S Whipple resigned from the Board because of the difficulty in travelling from London to Reading. They were persuaded to remain, but the result was that future wartime Board meetings were held in London at the Royal Institution.

The amount of work at the Institute's headquarters meant the cessation of all clerical work for the Physical Society. Nevertheless, amalgamation between the Institute and the Society did creep nearer when it was agreed to explore a proposal from the Physical Society for joint subscription. This was shelved, but the fact that it was considered showed the drift of thought. Relations between the Institute and the Society continued to be friendly despite the difficulties caused by the war.

During 1940 two distinguished physicists died. Each of them tendered much service to the Society and the Institute.

On 22 August 1940 Sir Oliver Lodge died:

> 'Sir Oliver Lodge began work in his father's business at 14 years of age. Through evening classes he was able to pass the matriculation examination and the intermediate examination in science with first class honours in physics. He then entered University College London, took a Doctorate in Science and was appointed the first Professor of Physics at Liverpool University. After being elected a Fellow of the Royal Society, he became Principal of the then new Birmingham University. He was President of the Physical Society from 1899 to 1901. He was drawn towards psychical research and through that into spiritualism with claims that he had experienced communication with his son Raymond, who was killed in the First World War. Sir Oliver had twelve children. He came from a prolific family; he was the eldest of nine and his grandfather had twenty-five children from three marriages'.

On 30 August Sir J J Thomson died at the age of 83:

> 'J J Thomson was, by general consent, the father of the electron. Joseph Thomson was the son of a bookseller and was a student at Owens College, which in due course was to become Manchester University. From there he became an undergraduate at Trinity College, Cambridge where later, and for twenty-two years, he was to be Master. He received the Nobel Prize for Physics in 1906. He was President of the

Oliver Lodge (President of the Society 1899–1901).

Physical Society from 1914 to 1916 and the second President
of the Institute of Physics from 1921 to 1923 and its first
Honorary Fellow. Like the Braggs, there was a father and
son affinity. In 1937 his son, Sir George Paget Thomson,
won the Nobel Prize for Physics, jointly with an American
physicist, Clinton Davisson'.

A wartime innovation in the Institute was the introduction of *Notes and
Notices*, effectively a wartime newsletter. It was thought that 'time
was not opportune for the issue of an applied physics journal', but in
the meantime the *Journal of Scientific Instruments* was expanded to
include, tentatively, topics on applied physics.

The war was not going well. The disasters in 1940—the withdrawal from France, the evacuation of Norway and the retreat from Greece—spread gloom in Britain. Perhaps a victim of one of these was a member of the Institute's staff, Lance Corporal W A Smith. Certainly he wrote in 1942 to the Board from a prisoner-of-war camp in Germany, thanking them for the food parcels he had received. In fact the Board was aware of the growing number of members in POW camps. Fees, of course, were waived, but sympathies went further in that it was resolved to send through their families 'technical text books and so forth'. One can sense an almost touching compassion in the Board's minute on this, reflecting a realisation of what was happening to such members far away.

By the spring of 1941 most large UK cities were being heavily bombed. Earlier, in the summer of 1940, a member of the Institute, R V Jones, was slightly bewildered to be called to a meeting set up by Winston Churchill to explain to the War Cabinet the technique German bombers were using to fly along radio beams for precision bombing. His success in combating the bombers caused Churchill afterwards gratefully to name Jones as 'the man who broke the bloody beams'.

By the end of 1941 a whole new field of cooperation was opening between British and American scientists after the attack on the American fleet at Pearl Harbor and the British declaration of war on Japan, a collaboration which would effect the defeat of Japan through Hiroshima and Nagasaki. The Board continued to pursue ideas on the best contribution physicists might make to the war effort. Dr F C Toy (to be President from 1948–1950) submitted a proposal that the Central Register be opened to graduates who had pass degrees to enhance flexibility and to tap into an undoubted well of ability at a time when physicists were greatly needed.

A further proposal originated from the Board. The shortage of physicists was causing concern and the Board suggested a simple solution. A short period of training would enable a physicist who was following a 'less immediate wartime application' to become a trained radio engineer. In other words, a release of students on degree courses to train for such work. Heavily emphasised within this plan was a guarantee of resumption and completion of university courses after the war for students drawn into electronic work, and a firm promise that

R V Jones was responsible for anticipating German applications of science to warfare and developing ways to counter these (1937 photograph from 'Most Secret War, British Scientific Intelligence 1939–1945' by R V Jones, © 1978 Hamish Hamilton).

students should where possible be trained in the subjects from which they were seconded. This carefully worded proposal led in part to the motion passed in the House of Commons:

'That this House is of the opinion that present circumstances require the early establishment of a whole-time Central

An early radar station during the Second World War (Science Museum/Science & Society Picture Library).

Scientific and Technical Board to coordinate research and development with relation to the war effort and to ensure that the experience, knowledge and creative genius of British technicians and scientists exert a more effective influence over the conduct of a highly mechanised war'.

The Parliamentary and Scientific Committee considered the statement. The Board's plan was put into effect and some students were seconded to electronic work.

On 12 March 1942 Sir William Bragg died. From this distance in time it is difficult not to regard the Braggs as twins rather than father and son. They shared the Nobel Prize for Physics in 1915; both were Fellows of the Royal Society, with the father President in 1935; both were Fullerian Professors of Chemistry at the Royal Institution; both were Directors of the Davy–Faraday Laboratory; both were knighted; both were among the first list of guarantors at the inception of the Institute of Physics; both were Presidents of the Institute, Sir William from 1925 to 1927 and Sir Lawrence from 1939 to 1943:

'Sir William was a native of Wigton, Cumberland where he was brought up on a farm. An uncle took over responsibility for his education and he won a scholarship to Trinity College, Cambridge. After gaining his degree he applied for and obtained the Chair of Physics and Mathematics at Adelaide University. During the First World War he worked on anti-submarine devices and later as a consultant to the Admiralty. He was President of the Physical Society from 1920 to 1922, and President of the Institute from 1925 to 1927'.

It was at this time in 1942 that the Institute published a recommended scale of salaries. This may make amusing reading today, but physicists, particularly younger ones at that time, were vulnerable and relied heavily on advice from their institutions as their support. As never before, the voice of the Institute mattered to them. The following figures were given:

New graduate with honours	£250 per annum
Graduate with honours after 5 years	£350 per annum
Age 30 and capable	£450 per annum.

No recommendations were made above £450 as no doubt other factors applied. It was stated at the time that the cost of living in London was £25 per annum, but it is not clear what that included.

Wartime secrecy and censorship, particularly on scientific matters, were quite ferocious and very well observed. As an example, a slightly ghostly conference on 'X-ray Analysis' was held at the Cavendish Laboratory in Cambridge on 11–12 April 1942. It was well attended by eminent physicists, but it vanished as it ended. The proceedings were declared confidential and no press account was published. In view of the involvement of members of the Cavendish in subsequent major wartime developments, one can speculate on the actual nature of this conference.

Some of the things being done by government were causing unease in some academic areas and the following exchange may be of interest. Students following science degree courses were being granted one-year deferment before being called up to some form of national service (on the grounds that their extra experience would be useful to the war effort). The headmaster of Winchester College wrote in *The Times* on 14 December 1942 of his anxiety over the Ministry's decision not to

Lawrence Bragg (President of the Institute 1939–1943).

grant arts students the same facility. He saw in this the end of ordinary arts courses in universities. He went on to say that there was a growing encroachment of technology on schoolwork, a growing demand for specialisation, pressures from fee-paying parents to have their children gain professional qualifications as soon as possible, and the influence of university scholarship examinations in science and mathematics. All this, said the headmaster, 'constitutes a real threat to the depth and balance of English education'.

The reply from the President of the Institute, Sir Lawrence Bragg, is worth quoting in full:

'Scientists will, I feel sure, warmly sympathise with many of the points made by the headmaster of Winchester in his letter to *The Times* today. We deplore the cessation of arts courses at the universities and, while realising that national interests have first claim, we hope with him that these courses will be restored at the earliest opportunity.

If I may speak for the science students who are allowed at present to continue their studies, I am confident that they regard this possibility not as a privilege to enjoy an education denied to their fellows, but as a preparation to play their part in a war which makes heavy demands for scientifically trained personnel.

There appears, however, to be a tacit assumption in his letter that courses in mathematics and science have a lesser claim to be considered a part of a balanced education than arts courses, that the former are 'technical' and the latter 'general' subjects. Here we join issue. Many of us deplore with him the growing tendency to specialisation and all its dangers. We agree that this specialisation has been forced upon schools by university scholarship examinations; there is an increasing conviction that the character of these examinations must be critically reviewed and radically altered in our educational plans for the future. Such specialisation is not confined, however, to science and mathematics, and can be equally undesirable in arts subjects.

Your correspondent pleads for 'those disciplines which seek to encourage free and well-informed thought upon great issues, religious, political, social and economic'. The most potent factor in bringing about our changed world in the last 100 years has been the influence of scientific thought and achievement upon religious, political, social and economic questions. These questions cannot be studied in a balanced way without some knowledge of scientific method. The fact that a 'general' education has so often in the past failed to include science has been responsible for many of the weaknesses in our national structure which the war has revealed. It is a tribute to the teaching of the literary subjects to admit that most scientists know something of the arts and

wish they knew more, whereas most of their arts colleagues
know nothing of science and are often rather proud of it; but
in so far as this is true it does not imply that the scientist has
had the narrower education.

In pleading for more study of arts subjects by those who will
later follow science, the headmaster of Winchester is beating
at an open door; enlightened scientific opinion will meet him
more than half-way. Cannot we also agree that the converse
is equally desirable, and that a knowledge of science is an
essential part of a balanced education?'.

The Arts and Sciences argument has raged since the 18th century, but
Sir Lawrence Bragg's succinct reassertion of the scientist's position in
the middle of the war was auspicious. It looked back to T H Huxley's
questioning in the 1880s of teaching classical education to the exclusion
of science. It found later echoes in C P Snow's controversial *The Two
Cultures* in which he said:

'A good many times I have been present at gatherings of
people who, by the standards of the traditional culture, are
thought highly educated and who have with considerable
gusto been expressing their incredulity at the illiteracy of
scientists. Once or twice I have been provoked and have
asked the company how many of them could describe the
Second Law of Thermodynamics. The response was cold:
it was also negative. Yet I was asking something which is
about the scientific equivalent of 'Have you read a work of
Shakespeare's?".

However the Board of the Institute still kept the future in mind.
H Lowery prepared a widely circulated leaflet, *Post War Education
and Training of Physicists*, and he was also active on a committee set
up by Lord Hankey to examine post-war education and resettlement.
A joint committee was also set up by the Institute and the Chemists to
pursue common interests with an emphasis on 'the use of the scientist
to the best advantage in the service of the community' and various
ideas were promoted. One concept was that, since the Central Register
had been so successful during the war, it might be common sense to
continue it afterwards.

Early in 1943 Sir Lawrence Bragg, still President of the Institute, was again writing to *The Times,* this time on the position of the scientist in society:

> 'Most scientists find themselves in agreement with the pleas which are now being made that the scientific effort, so effective in the war, shall not be allowed to cease when peace returns, but shall be applied to the serious problems which will arise during reconstruction and after. It is certain that if, in this small island with its limited material resources, we are to maintain our existing population with its present standard of living, scientific and technological research will be required on a scale not yet envisaged.
>
> Many responsible scientists, however, view with concern exaggerations which often accompany reasonable claims. From time to time statements are made by individuals or by organisations professing to speak for science, that if some fraction of the national income were allocated to scientific research, and if men of science were given a position of authority in the affairs of state, the community would find itself in what is usually described as 'an age of plenty'. It is unfortunate that such exaggerations should be disseminated when schemes for future re-organisation are being discussed. To mislead the community as to its available resources can only foster illusions and bring disappointments which may be disastrous both for it and for science. While we may hope that the improvement in our material comforts, which has marked the past 50 years, will be continued by further applications of scientific methods, the fruits of research sometimes ripen slowly and our material resources during the post-war period cannot be vastly greater than we now possess. Because of the time-lag in the application of research, it is important that immediate preparation be made for reconstruction.
>
> The claim that the scientist, as scientist, is entitled to some position of exceptional authority in deciding the policies of governments, is one that cannot and should not be accepted in a democratic community. Social problems are too complex to be solved by any one type of mind. The man of science can give valuable assistance in solving problems facing a

The Clarendon Laboratory, Oxford, during the Second World War: airborne infrared detectors were developed on the top floor of this building (from 'Most Secret War, British Scientific Intelligence 1939–1945' by R V Jones, © 1978 Hamish Hamilton).

society by searching out the facts and, on the basis of the facts, suggesting remedies. He could profitably be consulted more frequently than has been the case. When however his advice has been given, his duty as a scientist is at an end. No social problem can be solved solely by the methods of science; not only material but other values are involved, and it is for the community, of which the scientist is a member, to weigh the different factors and make a decision. A scientific and soulless technocracy would be the worst form of despotism'.

This *cri de coeur* was also signed by the President of the Royal Institute of Chemistry and the Chairman of the Joint Council of Professional Scientists.

It looked as if British scientists had by their contribution helped to nullify conventional enemy bombing. The German scientists, however, had two more weapons available, the V1, the 'flying bomb', and the V2, a rocket with a one-tonne bomb load. London and the South East were the targets and the destruction was great, but quite random. The Board, however, continued to meet at the Royal Institution though it was surrounded by a great deal of damage. The Board was busy. Talk of a Royal Charter came to nothing since no Royal Charters were granted during the war. Physicists in Scotland were showing interest in their own affairs and a branch was formed in 1944. China was showing interest in Institute affairs and in March 1944 the British Council asked the Institute for microfilm copies of the *Journal of Scientific Instruments* for circulation in China and the Board was happy to agree.

Slight embarrassment was felt at the Board's meeting in May 1944, the minutes of which recorded that 'the Board noted that, although the Australian branch had been active for some sixteen years, it had never been formally created and the Board now confirms the creation of the branch'.

With the end of the war in sight post-war matters began to loom large. With that in mind, Professor J D Cockcroft suggested a new journal for the Institute as he felt that the *Journal of Scientific Instruments* was not adequate as a main journal. This matter became one of concern and the Joint Committee of the Physical Society and the Institute accepted the dictum that all papers dealing primarily with original research were the concern of the Physical Society, and that papers dealing directly with scientific instruments were the concern of the Institute. The outcome was the proposal that a new publication with a joint editorial board provided by the Institute and the Society should be implemented in due course.

These deliberations were part of wider discussions taking place between the Institute and the Society on the difficulties in drawing dividing lines between the activities undertaken by each body. It was agreed that 'there would be everything to gain by a complete cooperation between the two bodies in certain common fields such as publications and groups'. Although it would still be another fifteen years before the amalgamation, the seeds were taking root.

The government announced a scheme in 1944 for a system of grants for servicemen returning to university courses at the end of the war, but the Board had its attention drawn to an anomaly in the scheme for

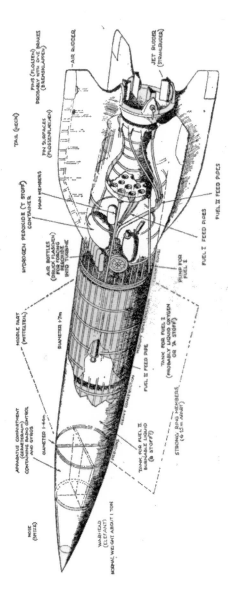

Fig. 30. Drawing by R. Lampitt of a V-2 before any fell in Britain. The outline was based on air photographs and on papers captured in Normandy. The nomenclature came from Enigma information, and the technical details overwhelmingly from the Farnborough examination of the remains of the V-2 that fell in Sweden on 13 June 1944. Later examination showed that the hydrogen peroxide container should have been below rather than above the pump

Artist's sketch of what the V2 rocket might look like inside (from 'Most Secret War, British Scientific Intelligence 1939–1945' by R V Jones, © 1978 Hamish Hamilton).

five Scottish students who were employed in the Armaments Research Department at the Ministry of Supply and needed grants on the same basis. It was typical of the Board to take the whole matter to review. It was already in discussion with the government about those students who had been seconded to radio work away from university. It was agreed by Clement Attlee, the Lord President of the Council, and by Ernest Bevin, the Minister of Labour, that they could go back into further education when it became possible.

As early as 1943, the Honorary Officers and the Council of the Physical Society were giving active consideration as to what would happen after the war. The future housing of learned societies was an obvious problem. On 22 December 1943 the President of the Society, Professor E N da C Andrade, wrote the following letter to *The Times* from the Society's offices at 1 Lowther Gardens:

'Sir: In his presidential address to the Royal Society, in which he dealt with the accommodation of the Royal Society and of the scientific societies in general, Sir Henry Dale said that, while the authorities of the 1870s did not see much future for chemistry, apparently they did not foresee any future for physics at all. No doubt he had in mind the Physical Society, which is at present housed, by courtesy of the Royal Commissioners for the Exhibition of 1851, in the top floors of what was a private house in Kensington, to which the only access is a forbidding flight of stairs. The Society is very grateful to the Royal Commissioners, who have always shown great kindness and consideration; without this help the Society would be in grave difficulties. At the same time, in view of the services which the physicists of the country have rendered to the State, it may be contended that it has deserved some worthier accommodation. The Society, which was founded in 1874, has increased its membership from about 100 in the early days to some 1350 at the moment, and it is growing rapidly. Among its past presidents it numbers Lord Kelvin, Balfour Stewart, G F Fitzgerald, Oliver Lodge, Arthur Schuster, J J Thomson, W H Bragg, Sir Arthur Eddington, Lord Rayleigh and Sir Charles Darwin. It would like to have a home of its own in a central position, near its great ancestor the Royal Society, and hopes that its demure virtues will not be neglected when the question raised by Sir Henry Dale comes up for serious consideration'.

In October 1944 Professor E N da C Andrade represented the Society on a delegation considering the future housing of scientific societies. This delegation, which was headed by the President of the Royal Society, met the Lord President of the Council, the Chancellor of the Exchequer and the Minister of Works and Buildings. The problem did not even begin to be solved by the meeting and for the Society adequate housing remained a pressing need for many years, as it was also to become a major concern for the Institute.

One positive development that took place during the war years was the establishment of groups of members concentrating on certain branches of applied physics. The first of these was the Colour Group inaugurated in February 1941. Membership of the Group was open and free to all Fellows and student members of the Physical Society. Members of other relevant societies, institutions and associations could join, but a payment of an annual subscription of 5s was required. Other persons could be introduced by a member of the Physical Society for an annual subscription of 10s. 6d. The group was an immediate success with an average attendance of 50 at the five meetings held during the first year, which was about the same as the average attendance at a meeting of the Society itself. Membership of the Colour Group was 132 at the end of its first year.

The Council of the Society decided to establish further groups similar to the Colour Group and in March 1942 the Optical Group held its inaugural meeting. In the first year of its existence it held five meetings with an average attendance of 50 out of a total membership of 303. Council recognised that the group was the best, if not the only, means of bringing together many who felt that their particular needs were not being met adequately by the merging of the Optical Society and the Physical Society of London. The Low-Temperature Group was inaugurated in October 1945 and the membership of the three groups at the end of 1945 was: Colour 178, Optical 326 and Low-Temperature 54.

CHAPTER 6

THE POST-WAR YEARS— APPROACHING THE MERGER (1945–1960)

8 May 1945 was VE day ('Victory in Europe' day) and so the war in Europe ended. Later in the year the first two atomic bombs were dropped and immediately ended the war in Japan. The klystron which had been developed for military purposes began to be exploited for peaceful use and the first microwave oven was on sale in 1947.

The Institute had made a tremendous contribution during the war: through its members, some of whom died in unrecorded circumstances, some suffering in prisoner-of-war camps, others whose job it had been to bring their scientific skills to bear on the technological struggle that warfare had become; through its headquarters, run by Dr Lang; through its Board members who gave enormously of their time and effort harnessing their profession into an effective war capability. Alongside this, the Board never lost sight of its contract to its membership and to securing the Institute's future after the war, and they did this with vigour. At the end of the First World War, when the Institute had its beginnings, physicists found themselves being classified as chemists. By the end of the Second World War they were assured of their place in history in their own right.

The Institute's return to London

Membership of the Institute had steadily increased during the war, largely owing to the number of Associates having almost doubled in number to 946. The number of Fellows, together with the Honorary

Fellows, had also grown and totalled 759. The Students numbered 491 and the Subscribers had increased to 335. The total membership was 2531 (whereas it had been 822 a decade before).

With the ending of the war, the Institute was anxious to return to London. It had approached the Royal Institution about leasing 19 Albemarle Street, adjacent to the Royal Institution. At first the Royal Institution had been reluctant to allow this, but eventually a lease was agreed and it was planned to move there on 30 July 1945. The Society arranged to continue in Lowther Gardens.

The move, however, was not to be seen by Major Phillips, who sadly died on 17 June 1945 after being Honorary Treasurer of the Institute for 20 years. He had played a prominent part in the establishment of the Institute and his influence had been felt throughout its existence and was to continue to influence future developments through his generosity towards the Institute in his will. It was to be his bequest which, 50 years later, contributed in a large extent to the move to 76 Portland Place.

The move took place to the small suite of rooms in 19 Albemarle Street and the first Board meeting held at the new offices was in November.

Changes arising from the end of the war caused problems and enquiries on professional matters poured in from members. One result was an Appointments Register being opened to advise members on job vacancies along the lines of the wartime Central Register. The Institute was asked for representatives on a variety of bodies, including the British Standards Institute, and also for help in providing examiners for examination boards.

But a crisis was at hand. The office had hardly settled in Albemarle Street before it was realised that the move had been a failure. The Board noted:

'We find that the accommodation rented by the Institute at 19 Albemarle Street is hopelessly inadequate for its work to be carried out efficiently and we have unanimously agreed that immediate action should be taken to rectify the situation'.

Exterior view of 47 Belgrave Square.

Immediate steps were taken to find a suitable post-war home for the Institute, fitted to its needs and appropriate to its much increased membership and its enhanced reputation, and in the last week of 1946 the Institute moved into grander and more spacious accommodation in 47 Belgrave Square.

Number 47 was the town house of the Earl of Mount Edgcumbe who, as Major Kenelm Edgcumbe, had become a member of the Institute in 1920. At the end of the war the Earl was faced with maintaining two properties, his country seat and this large house in London. He was an engineer and had been President of the Institution of Electrical Engineers in 1928. Although not a professional physicist, he had a

The Council Room, 47 Belgrave Square.

keen interest in physics and he invited the Institute to make 47 Belgrave Square its headquarters, for which he asked an extremely low rent of £710 per annum. The freehold of the house was held by the Grosvenor Estates, and the Earl's lease had a little over 30 years to run.

The building was in quite good condition, although some structural alterations were necessary and some war damage had to be made good. Little work was needed to adapt its ample rooms for office use, so the generous offer from the Earl was accepted. At last the Institute had a 'home' in the heart of London, commensurate with its growing standing as a professional body. Dr Lang, accompanied by the newly appointed Assistant Secretary, Norman Clarke, moved in with a predominantly new staff.

Four of the elegant rooms on the ground floor and first floor became the Members' Room, the Lecture Theatre, the Board (Council) Room and the Secretary's (Dr Lang's) office. The staff occupied the second and third floors, whilst the top floor (the former servants' quarters) became a caretaker's flat. Parts of the basement were modified to provide a staff canteen and a senior staff dining room

The house was the first unshared accommodation the Institute had enjoyed and its pleasure was evident in a letter to Fellows and Associates which began:

> 'As you already know the Institute is now in occupation of a stately home; actually it is scheduled as a monument. ... The internal fitments, however, are of the plainest, and much is lacking ... . Donations in kind or money, however humble, will be equally welcome, and no lists of donations will be published. ... May we point out that the large rooms can well accommodate massive pieces of furniture which it is often difficult to use in private houses'.

Generous gifts made redecoration possble. Brown Firth Laboratories provided carpets and upholstered chairs to match the presidential chair presented by the South Wales Branch. GEC Ltd provided the Board Room tables.

During the next decade the building remained much as it was in 1947. Not all the fireplaces were blocked up and, despite the central heating, Norman Clarke, the Registrar and the Assistant Secretary, still had a coal fire in his office. There was, of course, no lift but those who worked on the third floor did not complain. It was the porters who bore the brunt of the stairs as they carried up the reams of paper for the duplicating machine, which was on the third floor, and then carried it all down again, printed and stuffed into envelopes, for mailing to members. The first lift, for 'goods' only, was installed in 1963, but by then a Reprography Department had been established in the basement. The passenger lift came later, but the Secretary, Dr Lang, scorned its use as he mounted the stairs two at a time.

There was a great deal of activity by the branches and groups as the freedom from wartime restrictions came. New groups were formed, namely the Electron Microscopy Group and the Stress Analysis Group (now the Stress and Vibration Group). The Acoustical Group was formed under the auspices of the Physical Society and was welcomed by the Institute. An Industrial Atomic Energy Group was discussed and the Physical Society was also considering a nuclear physics group, but there was some concern about the political activities of a newly formed Atomic Scientists Association and so the matter was rested.

The war had exhausted Britain. Economically the country was virtually bankrupt and an abrupt end to the United States' Lend–Lease Programme meant that austerity and rationing would become even more stringent than in wartime. Food, tobacco and petrol were rationed, as was paper—which would affect the Institute. Various steps were taken to cope with the shortage: contributions of 2000–3000 words of academic interest were only to be published in summarised form; lighter paper was used; advertising space was rationed; all blank space was to be used, even if it meant no longer starting papers at the top of a page; outside covers were used to handle advertisements, so saving four pages. This was especially frustrating at a time when intense discussions were still going on between the Institute and the Society over the question of establishing a new journal in applied physics. It became clear that, for the time being, any new such journal was a practical impossibility. The Editor of the *Journal of Scientific Instruments* was authorised to include up to 40% of 'non-instrument' material, and from January 1948 the journal became the *Journal of Scientific Instruments and of Physics in Industry*.

The Physical Society and the Annual Exhibition

The Physical Society entered the post-war period in reasonably good heart and it continued to operate from 1 Lowther Gardens. The great importance attached by the Society to its Annual Exhibition of Scientific Instruments and Apparatus is shown by the alacrity with which it resumed the series in January 1946. The three-day exhibition, the 30th in the series, was held at Imperial College and opened by Sir Stafford Cripps, the recently appointed President of the Board of Trade in the new Labour government. The war had only been over for a few months and the conditions for the preparation and organisation of an exhibition could hardly have been less favourable to the exhibitors or to the Society's small staff. Nevertheless, despite all the shortages and difficulties, the high reputation established by the pre-war Exhibitions was well maintained. The attendance was unexpectedly large—almost 16,000, nearly twice as large as the attendance of 8700 in 1939. This enthusiasm may be put down to the optimism of the British in the early years of the peace that followed the long rigours of the war.

The cloud chamber at Manchester in which Rochester and Butler found evidence for unstable elementary particles (1947) (Science Museum/Science & Society Picture Library).

Since its inception the Annual Exhibition had been held in early January but the winter of 1947 was so extremely harsh that a change of date from winter to spring proved necessary. This change turned out to be so very convenient both for exhibitors and visitors that it was retained for the next 11 years up to 1958, showing that external factors can sometimes induce beneficial changes. The Exhibition was extended from three to four days and this allowed conditions to be made much more comfortable through the controlled distribution of dated and numbered tickets. The total attendance was again high at over 8000. The catalogue for the Exhibition, the *Handbook of Scientific Instruments and Apparatus*, had long been widely accepted as a valuable reference work both in Britain and overseas, where sales were strong. In this difficult year, 1947, its preparation was seriously interrupted by the national fuel crisis and the electrical shutdown early in the year, and no copies were available in time for the Exhibition though substantial numbers were sold subsequently.

The Annual Exhibition continued to be a prominent feature of the Society's activities. For 38 years up to and including 1954, the Exhibition, normally three days in length but sometimes extended to five or six days, was held at Imperial College. From 1951 to the end of the period the total attendance for each Exhibition topped 10,000 and reached a maximum of over 20,000 in 1953. It is amazing that the amount of organisation required for an Exhibition of this size could be accomplished within the compass of the small staff of the Society and clearly much reliance had to be placed upon voluntary help. In 1954 expansion of work at Imperial College meant that less space was available than in previous years. The Exhibition on this occasion was restricted to about 75% of its former size and the Society was informed that it would not be possible for future Exhibitions to be housed there. In 1955, therefore, the Society held its 39th Exhibition by hiring space in the Royal Horticultural Society's halls and this venue was used for many years. The change of venue did not seem greatly to affect the total attendance, which remained 16,000–18,000 until the amalgamation of the Society with the Institute of Physics in 1960.

The conferences and meetings organised by the Physical Society

This same spirit of optimism which pervaded the Physical Society after the end of the war and which had led to the speedy revival of the Annual Exhibitions may perhaps also be seen in the early organisation of an international conference by the Society in July 1946. By arrangement with Sir Lawrence Bragg, the conference on 'Fundamental Particles and Low Temperatures' was held at the Cavendish Laboratory, Cambridge:

> 'About 300 attended, of whom 100 came from abroad, representing Belgium, Brazil, China, Denmark, Egypt, France, Holland, India, Italy, Jugoslavia, Norway, Persia, Sweden, Switzerland, USA, USSR as well as the Dominions'.

The Holweck Medal.

The attendance was so large that many members of the Society who wished to attend could not do so. Professor Niels Bohr gave the opening address. One day was devoted to the work of the Cavendish and Royal Society Mond Laboratories and to the ceremonial opening of the Austin Wing.

There was an extensive programme of other meetings. During 1946 there were 11 general meetings in addition to 15 meetings held by the Colour, Optical and Low-Temperature Groups. Six of the general meetings were held in the Science Museum, one at Imperial College, three at the Royal Institution, and one in the Physics Department of the University of Birmingham. The Birmingham meeting was the first of a new series of meetings to be held at universities outside London.

The Holweck Prize

During 1945 a fund had been collected from Fellows and friends of the Society to finance an annual prize in the interests of closer cooperation with physicists in France and as a memorial to Dr Fernand Holweck, the distinguished Director of Research at the Institut du Radium in Paris, who was tortured and died at the hands of the Gestapo, and to other physicists who met their deaths or suffered great privation during the German occupation of France. The original idea was to award the prize annually for a period of ten years, alternately to a French and a British physicist for distinguished work in experimental physics. The presentation to the French winner would be made in London and to the British winner in Paris. The Société Française de Physique agreed to produce a Holweck Medal to be presented to each recipient of the Prize, initially bronze and now in gold. The first award of the Holweck Prize was made to Professor Charles Sadron of the University of Strasbourg in 1946. The Award has continued ever since and has already been received by more than 50 recipients.

The Physical Society and its publications

Publishing continued to be a conspicuous feature of the Society's activities. The three main publications were the *Proceedings*, *Reports on Progress in Physics* and *Special Reports*. But the effects of the war were still being felt, in particular the rationing of paper. A new section called 'Letters to the Editor' was introduced about this time and it proved so popular that it became a regular feature of the *Proceedings*. These Letters were not intended in any way to be substitutes for detailed publication of completed work but rather for preliminary announcements of results which might be of some importance to future work either by the author(s) of the Letter or others in the physics community.

The sometimes excessive delay which occurred between the submission of a paper and its subsequent appearance in the *Proceedings* was a recurrent concern for the Council of the Society. In January 1949 an attempt was made to meet this problem by issuing the *Proceedings* in two volumes with each volume consisting of six parts issued monthly. It was initially proposed that one volume should have papers dealing with pure physics and the other with technological aspects. This view was rejected as being an impractical method of division since papers on some topics might need to appear in both volumes. The Council felt that physicists would prefer all papers on a given topic to appear in one volume whether they had been written with applications in mind or not. So it was agreed that Section A of the *Proceedings* should cover atomic and sub-atomic physics and Section B should deal with macroscopic physics. Each member of the Society would receive one volume free and could purchase the other at a reduced price. To have supplied both volumes free to all members would have required the annual subscription to the Society to be raised to over four guineas and this was very firmly rejected by the membership at a Special General Meeting.

Despite rigid economies, the cost of publishing the *Proceedings* and of supplying *Science Abstracts* continued to rise inexorably and in January 1953 the Council introduced a new scheme for subscriptions in which subscriptions for publications were entirely separated from subscriptions for membership. The basic subscription for membership was reduced to two guineas but with no free publications, although members were offered preferential rates for the *Proceedings* and *Science Abstracts*. In addition, provision was made for the purchase at a nominal rate of separate copies of papers published in the *Proceedings*. This Reprint Service, which was a radical departure from established practice, did attract some initial interest but for various reasons had to be discontinued after three years. The question of the costs of the publications continued to prove a worry for the Council and in December 1957 it was decided that the two sections of the *Proceedings* should once more be issued as a single volume.

The annual book *Reports on Progress in Physics* was one of the most important services rendered by the Society to the physics community. These annual reports started in 1934 and were issued annually until 1942 when Council had to combine two years in one volume, although the use of two years' supply of paper into a single volume meant that it was possible to print a larger number of copies than in previous years. This was just as well because there had been a particularly large demand for the book in this country and overseas. The combination of two years in one volume continued in 1944 when Council commented that:

'While physicists remain so busily engaged on work of the highest importance for the Allied cause, and while the supply of paper remains so meagre, it has again been decided that the next volume must, like its immediate predecessor, cover two years instead of one'.

Physicists returned from the war in 1945 but paper supplies remained meagre so the combination of two years in a single volume continued until 1950 when the publication of *Reports on Progress in Physics* reverted to an annual basis.

The third aspect of the publishing work of the Society, the publishing of *Special Reports*, remained relatively much smaller than the other two. Some early examples were *Report on Defective Vision in Industry* (1946), *Meteorological Factors in Radio-wave Propagation* (1947), *Report on Colour Technology* (1948) and *The Emission Spectra of the Night Sky and Aurorae* (1948). In general, the sales of the *Special Reports* were disappointing and at the end of 1948 Council decided that in future the production of such *Reports* would be kept to a minimum, essentially to reports of some of the conferences in which the Society had a particular interest.

The finances of the Physical Society

During the war the membership of the Society had grown to a high point of 1457. This trend was good, but unfortunately there was also a large financial deficit and the Report of Council for 1946 commented that this deficit was 'a sharp reminder that the present rates of annual subscription are far too low in comparison with the services rendered by the Society'.

Although membership continued to grow, the financial position deteriorated further and was addressed at two Extraordinary General Meetings in 1947. At the first an incremental scale of composition fees for life membership was adopted. At the second a special resolution to increase the basic annual subscription of Fellows from two to three guineas and of student members from 10s 6d to 15s 0d was adopted. A third such meeting illustrated the innate conservatism of the Fellows: a proposal to replace the *Proceedings* by a magazine led to an extended discussion and the proposal had to be withdrawn because of the fierce opposition of so many Fellows.

The financial situation of the Society continued to be an uneasy one. It is true that in many years income was above expenditure, even if the excess was usually very small. The Society, however, had to rely heavily on hospitality in London, rent-free rooms for meetings and—even more important—free accommodation for the Annual Exhibition. Almost every Annual Report had to express the Council's gratitude along the lines of the Report of Council for the year ended 31 December 1950:

'The Council again records the cordial thanks of the Society
to the Rector and Governing Body of Imperial College and
Professor Sir George Thomson and Professor H V A Briscoe
for the great privilege of holding the Exhibition in the Physics
and Chemistry Departments of the College. For the use of
the Lecture Theatres for Science Meetings of the Society and
its four Groups the Council thanks the various bodies who
gave them hospitality, in particular Sir George Thomson,
Head of the Physics Department of Imperial College, the
Director of the Science Museum, and the Managers of the
Royal Institution'.

A quite serious problem arose in 1951 when the Royal Commissioners
for the Exhibition of 1851, who for many years had provided the Society
with free accommodation for its headquarters at 1 Lowther Gardens,
found it necessary to ask for a rent of £500. This was, of course, a
moderate sum for the space occupied by the Society, but it was still
worrying for a Society which in the year ending 31 December 1951
had a balance of income over expenditure of only £163.

In 1956 the Society's accounts were presented in a new and more
informative way. The income and expenditure of the three main
activities of the Society were shown as separate accounts, allowing
members to see much more readily which activities cost money and
which brought income to the Society. In 1956 the figures were:

	Income	Expenditure	Balance
Membership Services	£ 4866	£ 5582	−£ 716
Publications	£27309	£28969	−£1660
Exhibition	£15823	£12419	£3404
Totals:	£47998	£46970	£1028

The figures for 1957 were similar in terms of the proportion of resources
allocated to the three activities though a negative balance of £3179
for publications brought the overall excess of income over expenditure
down to a mere £20. This led the Council for the first time in many
years to raise substantially the price of the Society's publications sold
to libraries worldwide. This had an immediate effect and produced

a positive balance for publications in 1958 and 1959. Membership services continued to have a negative balance, whilst exhibitions had positive balances in 1958 and 1959. The result was that the overall balance was £4581 in 1958 and £3796 in 1959.

The above shows that the financial position of the Physical Society was not such that the future for the Society could be faced with great confidence. This was certainly one of the factors leading towards the amalgamation which was to come. But before we come to the story of the amalgamation, we should look back at the Society and take note of what the Institute was now doing.

Looking back at the Physical Society

Perhaps the most remarkable fact about the first 80 years of the Society's existence is the extent of its activities. The number of meetings and exhibitions organised, the number of papers read, the number of demonstrations given and the amount of material published were all prodigious, particularly when it is appreciated that most of the work had to be done by Honorary Officers and members. It is interesting to look at the Treasurer's Reports over the years. For example, in 1900 the total annual expenditure was about £660 and £550 of this went on printing the *Science Abstracts* and the *Proceedings*. In 1920 the expenditure was about £1370, in 1930 about £4470, and in 1940 about £3450. Only a very small fraction of these expenses refers to 'Secretarial and Clerical Assistance', and this makes clear that the Society relied heavily upon the goodwill of its membership. It is little wonder that it was viewed affectionately by its members, many of whom did not want to lose something of its spirit through amalgamation.

In the first half of the 20th century, the development of the Physical Society had been remarkable. The membership had continued to rise; the status of the Society had increased enormously, and many eminent scientists were closely associated with it; prestigious lectures and meetings had been held. The Society's Annual Exhibition had developed in such a way that up to 9000 people had attended many of them. Then the country was plunged into war in September 1939. The war years were inevitably difficult for the Society, but it had managed to survive and its publications had been printed. In the post-war era it had concentrated on the activities which had been significant in the pre-war years, namely its annual meeting, its conferences and meetings, and its publications. But it was becoming increasingly difficult to identify the dividing line between some of its activities and those of the Institute and it was inevitable that there would be more and more discussion about a merger if what the Society stood for was to survive. Before we take this story further, we should look at what was going on in the Institute during these post-war years.

The work of the Institute in the post-war years

Once the Institute had moved into 47 Belgrave Square, it became more and more involved in an ever-increasing range of activities.

Membership

Regulations for membership of the Institute were reconsidered in November 1948. A broadening of Fellowship requirements was recommended as it was felt that the emphasis on the DSc degree as a guide to standard was narrow and quite inapplicable to senior physicists who had not had the opportunity of academic research.

The new grade of Graduate was proposed, degree-based with a minimum age of 21, and this was introduced in 1949. 54 members in the first year became 283 in 1950, and continued to grow after that. Plans began for a Graduateship examination as an alternative means of admission and this was establshed in 1952. Written papers were taken in either Belgrave Square or the Mining and Technology College in Wigan, with the practical examination at the Regent Street Polytechnic in London. Of 21 candidates, six passed. The number of candidates was to increase greatly in the following years. The Board was gratified by the decision by the Ministry of Labour to grant Graduateship students deferment from call-up for National Service.

A number of goals were achieved in 1956: the 1000th Fellow was in membership, the 2000th Associate and the 1000th Graduate member; 1018 Fellows and Honorary Fellows, 2008 Associates, 1091 Graduates, 461 Subscribers and 568 Students were on the register. But there was no lowering in effort to increase membership and during 1956

The ACE (Automatic Computing Engine) 'pilot' model computer at the National Physical Laboratory, 1950 (Science Museum/Science & Society Picture Library).

Bulletins were sent free of charge to professors of physics in universities throughout the country as a gentle recruitment reminder.

There had been one decrease in membership: in March 1950 Council decided that it would have to take measures to expel Klaus Fuchs from membership on the grounds of professional misconduct, but Fuchs resigned his Fellowship before this could happen. Previously a student of Kiel and Leipzig, he emigrated to Britain in 1933 and went to Bristol and Edinburgh Universities. He was naturalised in 1942 and became a Fellow of the Institute in 1944. He was part of the team which Sir Rudolf Peierls took to the United States to work on the first atomic bomb, the Manhattan Project. On his return to England he worked at Harwell until 1950 when he was arrested for spying and passing secrets to the Soviet Union. This began a string of security scandals involving Guy Burgess, Donald Maclean, Kim Philby and later Sir Anthony Blunt. Fuchs spent fourteen years in prison.

Branches and groups

The work with branches and groups was growing fast. The North East Coast Branch (the future North Eastern Branch) was formed in 1949 and the newly formed Education Group held its first meeting in October of that year.

At the end of 1949 the branches and groups had the following membership:

Australian Branch	348
Indian Branch	60
London and Home Counties Branch	1540
Manchester and District Branch	564
Midland Branch	320
Scottish Branch	185
South Wales Branch	176
North East Coast Branch	145
Education Group	206
Electronics Group	298
Electron Microscopy Group	190
Industrial Radiology Group	200
Industrial Spectroscopy Group	110
Stress Analysis Group	150
X-ray Analysis Group	300

There was increasing activity throughout the branches and the groups as recovery from the war took hold. This resulted in an annual conference of branch and group officers. In May 1947 one of their submissions urged 'the Board to consider whether a new body should be formed, either by physicists alone or by physicists in conjunction with members of other professions, to give greater protection to the professional and economic status of its members than is possible through their present professional organisation or organisations'.

While this stopped short of a suggestion for a trade union, it did emphasise to the Board of the Institute that members would continue to regard their Institute as being much more than a London club. In fact, the status of physicists was under almost constant discussion, as were salary scales, and there were several meetings with the Parliamentary Scientific Committee on relevant matters.

A branch for Liverpool and North Wales held its inaugural meeting in May 1955. It quickly attracted 160 members. Group officers felt that there was interest, particularly among non-physicists, in joining single groups and a new grade of 'Group subscriber' was accepted on trial at an annual fee of one guinea. Group subscribers were not regarded as members of the Institute.

Overseas branches

Discussions with physicists in Australia had begun as early as 1923. There was enthusiasm amongst them to organise themselves into a coherent community with strong links with the Institute. By 1928 the number of physicists and the number of activities had justified branch status. In 1953 the Australians circulated their membership and found that the majority were against branch status, but concluded that they were not yet ready for complete independence. The Board in London was willing to accept whatever the Australians proposed and its policy was to help the Branch to move towards independence. It was during 1962 that the Australian Branch finally became an independent body, the Australian Institute of Physics. Special arrangements were made to enable Australian members to continue to belong to both their own Institute and the Institute in London at a reduced rate. The Council of the new Institute put on record the debt it felt for the encouragement and assistance it received and it made a presentation of furniture, made from Australian wood, for the Members' Room in Belgrave Square.

In the same year that the Australian Institute of Physics was formed, a group of members in New Zealand decided to form a New Zealand Branch and in 1982 this too became an independent New Zealand Institute of Physics.

In India a committee was formed in 1931 to deal with membership applications to the Institute and in 1934 an Indian Branch was formed under the chairmanship of Sir C V Raman. However, in 1955 Raman suggested that he should retire and the Branch should be reconstituted. The Board hoped he would continue in office and could find no justification to change the Branch rules. But within a year, Raman wrote:

'If I understand the mentality of my countrymen aright, the Fellowship of the Institute of Physics is valued just because it represents recognition originating outside India. In the circumstances, it is appropriate that the applications should go directly to the London office of the Institute. I have never been convinced that the existence of the Indian Branch of the Institute serves any useful purpose and would unhesitatingly recommend that it be closed down'.

In consequence of this, the Indian Branch was dissolved forthwith.

One other overseas branch was formed in the post-war years: a branch in Malaysia to cover the interests of physicists in both Malaysia and Singapore. However, in 1973 it became illegal in both those countries for there to be a branch of an organisation based outside national boundaries. Thus there was no alternative but to dissolve the Institute's branch in Malaysia. A new independent Malaysian Institute of Physics (Institut Fizik Malaysia) was formed within Malaysian boundaries, but not including Singapore, which was now a separate country.

Professional matters

One responsibility of the Institute was to publish a salary survey for the guidance of members. The 1948 survey showed:

Associate at 30	£600 per annum
Associate over 40	£750 minimum per annum
Fellow 30 to 40	£650–£1250 by experience and ability.

Another of the Board's concerns was conditions in physics laboratories. It felt the matter urgent enough to set up a special committee which included the Secretary of the Royal Institute of British Architects and Building Research Station representatives.

Much concern was felt within the Board that there were too few physicists to fill the posts now available and this question of supply and demand of physicists was a recurring theme throughout the fifties.

Members were turning to the Institute for help with their situations in industry, particularly in sensitive areas. One Fellow suspended from the Atomic Energy Research Establishment at Harwell on account of alleged 'unreliability' approached the Institute for help. The President discussed the matter with Sir John Cockcroft, who approved of the Institute's involvement and authorised it to state that the Fellow concerned was a 'competent physicist and an admirable technician'. The Fellow concerned quickly found another job, but this episode underlined that security was an issue of concern to members of the Institute.

For the Institute one of the more important post-war bills to pass through Parliament was the 'Radioactive Substances Bill' of 1949. Four Fellows, W Binks, Sir John Cockcroft, W V Mayneord and Sir George Thomson, were appointed to serve on the Statutory Advisory Committee and were asked to 'bear in mind the interests of experimental physicists'.   The newly formed Institute of Biology extended an invitation to participate in a meeting on 'Radiation Hazards' which the Board viewed with some reservation. Sir John Cockcroft attended, but not as a representative of the Institute. The issues involved were having practical effects on members of the Institute. One Associate member was being charged an extra premium on his life assurance because of supposed hazard. The Institute took up the matter and the additional premium was withdrawn and members were informed of what had happened.

Protection from radiation hazards was still under discussion. The Chief Inspector of Factories had drawn attention to the 'appalling ignorance' about radiological protection which he found after a serious accident concerning radiation and went on to say, disquietingly, '... in my view, firms should not be allowed to have x-ray equipment or radioactive materials unless they know of the dangers'. The Institute stressed in the *Bulletin* that proper precautions were essential when working with radioisotopes and x-rays.

In 1956 Russian troops invaded Hungary and put down a popular democratic uprising.   There were resulting refugees and requests were made to the Board to help place displaced physicists.   This work was willingly undertaken and the Deputy Secretary interviewed the physicists concerned and then contacted appropriate Fellows for help.

Education

The Institute was involved in educational schemes concerning the National Certificates and Laboratory Arts Certificates and had issued reports on education and training. It associated itself with summer schools and refresher courses in physics.

During the mid-50s the shortage of science teachers in schools was causing concern as it continued to do for the next decades. Inadequacy of salary was considered a reason, but also poor prospects as the best physics teachers could advance neither fast enough nor far enough.

The Institute became much involved in all aspects of physics education in secondary schools and this involvement is described in chapter 9.

Publications

From January 1950 the 'Physics in Industry' section of the *Journal of Scientific Instruments* became a new journal, *the British Journal of Applied Physics*, and a monthly *Bulletin*, forerunner of the present *Physics World*, was to replace *Notes and Notices*. The monographs in the *Physics in Industry* series were a proven success and two further volumes were planned. The evolution of the publishing activity is described in chapter 8.

Conferences and meetings

The number of meetings and conferences continued to grow as the branches and groups were arranging their own programmes.

1947 was the jubilee year of the discovery of the electron by J J Thomson and celebrations were held jointly with the Physical Society. Lectures at the Royal Institution and the Institution of Electrical Engineers drew crowds of 500 while at the Central Hall, Westminster, 2500 attended. The exhibition at the Science Museum was an undoubted success, went on until January 1948 and drew a public of 143,000. A sign of the austerity of the times was a feeling among members that a proposed dinner at the Savoy Hotel should not take place.

The Annual Conference of Branch and Group Officers for 1949 took place in Buxton during May, and it was immediately followed by the first Convention of the Institute, primarily for members and their guests. It was attended by about 10% of the membership.

The second Convention was held in Bournemouth in May 1953; some 450 members attended and the two main lectures were by Dr A C Menzies and Sir John Cockcroft. An evening lecture on 'The impact of physics on science and religion' was given by Canon C E Raven. The Institute's third Convention was held in Oxford in 1957. This attracted a surprisingly low number of 250 members and guests compared with previous conventions. A *post mortem* discussion of the Board decided that the next Convention should be held in an 'attractive seaside town' where accommodation would be easier.

One of the most successful conferences of the Institute was held in July 1956 in London on 'The Physics of Nuclear Reactors'. Its purpose in keeping members of the Institute and sister institutions abreast of development in nuclear reactors was appreciated by an attendance of 500 from 25 different countries.

The Institute had, of course, held dinners at its conferences and other functions, but it was not until March 1958 that the first Annual Dinner was held. There were 267 members and their guests present at the Savoy Hotel and the main toast to the Institute was proposed by Viscount Hailsham. However, at 35s a head (£1.75) it was apparently considered a bit costly and the following year it was held at the Park Lane Hotel at 27s 6d each (£1.40).

1951 was a mixed year for the country. In the midst of continuing post-war austerity, with meat rationing at its lowest level, the Festival of Britain was an enormous success. 27 acres were cleared on the south bank of the Thames and a 'Dome of Discovery' with the Royal Festival Hall and a 'Skylon' were built. Professor Tyndall represented the Institute on the Physics Advisory Panel. The journals were on an exhibition stand in the Dome and brought many aspects of science and technology to the notice of the general public.

A reception was given in 47 Belgrave Square by the Earl of Mount Edgcumbe on the eve of the Coronation in 1953.

Organisation

There were problems caused by the increasing workload and Dr Lang sent a memorandum to the Board in late 1948 asking for instructions regarding the priority to be given to jobs. His staff of 16 was barely adequate and unable to take on additional work. Note was taken by the Board and further assistance was given to him. The post of Deputy Secretary was created and Norman Clarke was appointed to it in 1950.

The Appointments Register set up after the war seems not to have flourished and was suspended in January 1950. Details of posts on offer were now covered by the *Bulletin*.

Dr Lang was trying to reduce the administration. The panel of consultants which had been set up after the war was not working. It was unwieldy with thousands of names registered and less than a dozen requests a year for consultants. The arrangement was stopped in 1951.

The position of the Participating Societies was re-examined. The original scheme for federation and integration of activities had not taken place and their position was seen as anomalous. As a result, in 1953 the Faraday Society, the Royal Meteorological Society and the British Institute of Radiology withdrew. The Physical Society alone remained of the original participating societies, but of course was unique among them in that it was the Society on whose initiative the Institute was founded in 1918. The Society continued to have two seats on the Board of the Institute.

Minor financial crises which did occur for the Institute from time to time led to an increase in subscriptions from January 1950 and these were readily accepted by the membership.

Ethical issues

During the war invitations had come from Robert Oppenheimer, the leader of the Manhattan Project, for British scientists to participate in the production of a nuclear device for military use. Rudolf Peierls, Sir James Chadwick, Klaus Fuchs and Joseph Rotblat were among the members of the Institute and Physical Society who went. The two bombs dropped on Japan finally ended the war, but anxiety was aroused in the lay public, as well as among scientists, on the devastating power now available. An illustration of the concern came from the Australian Branch in 1946. The New South Wales division had sent a letter to the Board complaining of 'conflicting' statements in the *Notes and Notices* distributed to members by the Institute in April 1946, which read:

> 'During the past months, suggestions have been received by the Board from members at home and overseas that it should express the Corporate view of the Institute regarding the use of Atomic Energy and how far the scientific and technical information connected with recent developments should be published. The Board is of the opinion that no useful purpose could be served by the issue of such a statement. ... It desires members to know that it is in full agreement with the views expressed by the President of the Royal Society'.

Sir Henry Dale, the President of the Royal Society, had said:

> 'If, by submitting for a while to secrecy, we could help to save that freedom and establish it for ever we could not hesitate; but we must be watchful now against any easy assumption that submission will be continued into peace. ... It is surely our duty as men of science to help the world with our knowledge to make that decision, and to make clear our own views and intentions. ... I think that we, as scientists, should make it clear to the world that, if national military science were allowed thus progressively to encroach upon the freedom of science, even if civilisation should yet for a while escape the danger of final destruction, a terrible, possibly a mortal wound would have been inflicted on the free spirit of science itself'.

Feelings in the Australian Branch were undoubtedly strong. Professor Ross, the President of the Branch, even wrote to the Board stating that some of their members—particularly the younger ones—felt that the Institute should express 'publicly and authoritatively its views on the situation created by the use of the atomic bomb and the suggestion for secrecy and control with respect to the use of atomic energy' and enquiring whether the Board would take action in this way.

John Adams, leader of the Proton Synchrotron construction team at CERN, Geneva, Switzerland, and its first Director-General (Science Museum/Science & Society Picture Library).

The Board could do little more than reprint the pronouncement of the President of the Royal Society, reiterate that it was in full agreement with that view and state that it strongly believed that atomic energy should only be used for peaceful purposes and that international agreement would be necessary to bring this about. The statement went on to say that 'since international agreements are obtained by political action, the Board as a scientific body does not feel called upon to express any opinion as to the steps required to realise such aims'. The Australians were mollified, but it does look as if the Board was caught on the wrong foot with regard to a very sensitive issue. Of course it also shows the influence attached to Board utterances by members, however far away.

In 1952 the first British atomic bomb was exploded in October on the Monte Bello Islands, off north-west Australia, which gave emphasis to the concerns voiced by the Australian Branch.

The Institute in the late 1950s

In the 1950s a breeder reactor was built, a bubble chamber was developed and the first fusion bomb was exploded in the USA. The concept of 'strangeness' was introduced in particle physics in 1953 and the following year saw the double-helix structure of DNA determined. CERN was set up, neutrinos were detected experimentally and antiprotons were produced. The BCS theory of superconductivity was propounded in 1957.

'Kitchen sink' drama arrived in London during 1956. John Osborne's *Look back in Anger* opened at the Royal Court Theatre and split the generations. Premium Bonds started with prizes up to £1000 and Britain's first nuclear power station was opened at Calder Hall. Sir John Cockcroft received a 'Royal Medal' from the Queen for distinguished work on nuclear and atomic physics. Russian leaders Bulganin and Khrushchev visited London.

In 1957 Herbert Lang completed 25 years' service to the Institute and a presentation was made to him. He had been propelled into the job amid slightly disquieting circumstances and had presided over the complete move of the wartime Institute to Reading. Organising its return after the war into the double shift from Albemarle Street to Belgrave Square must have been stressful and a toll was about to be taken. In the summer of 1957 he suffered a complete nervous breakdown. Hopes that it was mild were ended when he had a relapse and had to return to hospital. He was granted leave of absence. Norman Clarke took over the secretarial responsibilities until Lang returned in time for the March 1958 meeting of the Board.

Towards the end of 1958 the British Association for the Advancement of Science wrote to the Board to sound out the Institute's view on the adoption of a metric system of weights and measures. The letter was answered, expressing some surprise as a metric system was already the system normally used by British scientists.

What also seemed surprising was that the *Oxford Dictionary* still defined physics as the plural of physic. Dr Lang wrote to the publisher, gently pointing out that 'Nowadays this is not so and physics is a subject of its own. May I therefore invite you in future editions to consider giving physics an entry of its own, in a similar manner to chemistry and biology?'. The Secretary of Oxford University Press agreed with some alacrity to alter the entry.

Amalgamation

The question of amalgamation was never far from people's thoughts during these years. At a meeting between the respective Presidents of the Institute and the Society in late 1946 earlier ideas were hardened into a statement more definite than before. It began:

> 'Our discussions confirmed the view that we had previously reached independently that we are faced with a recurring danger of divided counsel which is not in the interest either of the profession or of the science which we serve. We personally believe, therefore, that the time is now ripe for a careful consideration of the possibility of amalgamating the two bodies'.

There were two issues: 'there must be a grade of membership in the new body for those Fellows of the Society who do not possess a professional qualification which would be regarded as adequate by the Institute' and the merger 'must avoid any suggestion that one body is to be absorbed by the other'. The Board decided that the views of groups and branches, at home and overseas, should be sought and the matter rested.

However, in 1947 the Board of the Institute was conscious of the 'protection' it could give the professional status of its members. It stated: 'we see no reason why the Institute should not undertake as much 'protection' work of its members as does the British Medical Association'. The Board did not approve at that time of some terms being offered in job vacancies. This issue raised concern about the proposed amalgamation which might prevent such 'protection' work in the future. A result was that a meeting of Branch and Group Officers of the Institute was held at which it was unanimously decided that all further discussion of amalgamation 'should be postponed and not be re-opened without full discussion with Fellows and Associates'. The decision of these officers effectively ended the possibility of amalgamation for a while.

Discussions about cooperation between the Society and the Institute did continue, nevertheless, in a somewhat desultory fashion over the years. In 1951 there was a series of meetings between the officers of the two bodies to explore the possibility of closer collaboration. It was agreed that members of each body should receive notices of meetings of the other, and should be entitled to obtain one of the journals of the other body at a reduced subscription rate, but such collaboration was at a restricted level.

It was a letter from Nevill Mott, President of the Society in 1957, which did have a greater sense of purpose than before. He suggested in this positive letter that the question of amalgamation be re-opened and he asked for a meeting of the Honorary Officers. The Board was conscious of the resolution unanimously passed in 1948 that 'no negotiations should begin until a full opportunity for discussion had been given to Fellows and Associates', but a joint meeting of Honorary Officers did take place and there was meaningful dialogue. Cogent reasons for amalgamation were put, including: 'young physicists would be more likely to join a single organisation as a matter of course, but are often bewildered by the claims of two'. At the same time, firm safeguards needed to be in place: 'in any form of amalgamation the status of the professional activities of the present Institute would need to be preserved'.

In 1958 a mutually agreed document *Memorandum to Members—Proposal to Amalgamate the Institute of Physics and the Physical Society* was circulated to members of both bodies.   There was substantial support from members and a Joint Amalgamation Committee was set up. Discussions about amalgamation continued for much of 1959 and in August of that year plans were circulated to all members setting out a general scheme for the complete amalgamation of the two bodies.

The original proposal for the name of the new body was 'The Society and the Institute of Physics', but this was soon changed in the discussion to 'The Institute of Physics and The Physical Society'.   Murmured objections were still being heard, mainly about what were considered inadequate safeguarding of the professional work of the Institute, the status of its corporate members and the constitution of the proposed new body.   Even so, and though it was to be a further two years before the amalgamation would happen, it was obvious by 1958 that the Board of the Institute and the Council of the Society intended that it should.

The two years leading up to the amalgamation were intensely busy. Joint meetings on the amalgamation were frequent throughout 1959. Fine tuning was being done and dates set for the formation of the new body and the winding up of the old. Dr Lang was to be the first Secretary after the amalgamation, with Dr Alice Stickland as Editor and Deputy Secretary and Norman Clarke as Deputy Secretary.

The broad basis of the scheme was outlined in a circular to members and was widely approved. Financial and legal details were sorted out. Useful comments and suggestions were absorbed. At an Extraordinary General Meeting of the Physical Society held on 27 November 1959 this scheme, with a few minor amendments, was approved by its Fellowship, though by no means unanimously. At an Extraordinary General Meeting of the Institute of Physics on 1 December 1959, the final scheme was similarly approved together with a request to 'the Board and Council to take the necessary action to implement the proposals for amalgamation'. It was agreed that the new body would be called 'The Institute of Physics and the Physical Society'.

Thus ended the separate existence of a Society founded 86 years previously, essentially through the efforts of one man, Frederick Guthrie. The 99 original Fellows had grown to a body of 2250 by 31 December 1959. In looking at the activities of the Society over its 86-year life the main reaction must be one of amazement at how much was accomplished with so little in the way of assistance from professional staff and financial resources. Its success as a learned society for physics was due mainly to the fact that the Society demanded and received great loyalty and service from its officers and from the membership in general.

CHAPTER 7

1960 AND ONWARDS

The decade of the sixties saw Britain looking back as well as forward. National Service, a relic of the war, ended. The millionth Morris Minor was produced and Britain applied to join the Common Market. Spies were still in fashion, both factual as George Blake and Kim Philby, and fictional with the first '007' books and films. The Beatles received MBEs and representatives from over 100 nations came to Winston Churchill's funeral in 1965. The majority, but not all, of the membership of the Institute and the Physical Society looked forward with enthusiasm to the prospect of bringing together the two bodies.

The first Council of the amalgamated Institute and Society was a 'caretaker' one appointed by the Board of the Institute of Physics and the Council of the Physical Society to hold office until 30 September 1961. It met as the Council-elect in January, March and May 1960. After incorporation of the new body under the Companies Act 1948 on 17 May, it commenced its regular meetings, taking over the work of the Board of the Institute and the Council of the Society, which bodies each held their final meetings on 5 July 1960. Thus was accomplished the merger which was first suggested in the 1920s and was repeatedly discussed in the intervening years as a logical development.

The first Council consisted of the President of the combined body, Sir John Cockcroft FRS; the Immediate Past Presidents, Sir George Thomson FRS (Institute) and J A Ratcliffe FRS (Society); the Honorary Treasurer, Dr J (later Sir James) Taylor; the Honorary Secretary, Dr C G Wynne; Secretary, Dr H R Lang; Deputy Secretary, Norman Clarke; and Editor and Deputy Secretary, Dr A C Stickland.

Although in the strict legal sense the original branches and groups ceased to exist with the ending of the two bodies, new ones had already

John Cockcroft (first President of the Institute of Physics and the Physical Society, 1960–1962).

been formed by the amalgamated body, so that in practice the work and activities continued uninterrupted throughout 1960.

The Institute membership before amalgamation was:

Fellows and Honorary Fellows	1205
Associates	2803
Graduates	1552
Students	988
Subscribers	410
Total	6958

The Physical Society at the time of the merger had:

Fellows and Honorary Fellows	1913
Students	390

a total of 2303, but 812 of these already belonged to one of the grades in the Institute.

At the end of December 1960, the combined membership was:

Fellows and Honorary Fellows	1215
Associates	2882
Graduates	1690
Total in professional grades	5787
Fellows of the Physical Society (not in professional grades)	1289
Students	1237
Subscribers	407
Total	8720

For some time before amalgamation a sub-committee of the Institute had been revising the qualifications necessary as academic requirements for the various grades. There remains to this day a category of members who prefer to be called 'Fellows of the Former Physical Society' rather than be classified as members of the Institute, but the number is steadily and inevitably declining.

There was one topic which kept on recurring throughout the 1960s, the name 'The Institute of Physics and The Physical Society'. It was felt by members to be cumbersome, and there was even greater distaste for the abbreviation IPPS. Despite this, Council decided not to propose any change for the present, even though it was an issue repeatedly raised by members. (Eventually it was finally settled when application was made for the Royal Charter in 1969 and only the name 'Institute of Physics' was acceptable to the Privy Council, but that was still some years ahead.) There was, however, agreement that the title 'The Physics Exhibition' was appropriate for the annual display of instruments and apparatus, inaugurated by the Physical Society in 1905.

Alice C Stickland (Deputy Secretary and Editor, 1960–1966).

During 1960 detailed attention was given to the reorganisation of the different departments. It was decided that the editorial work should be concentrated in the Society's rooms at 1 Lowther Gardens, South Kensington, under Dr A C Stickland. The Institute's house in Belgrave Square was to become the headquarters of the new body where all other activities were to be concentrated. Even as early as 1963 discussions were held as to whether new premises should be found to replace 47 Belgrave Square and the rooms in Lowther Gardens, but the cost of the various moves which were proposed were all too high to be acceptable. Council therefore agreed to some alterations in Belgrave Square as an interim measure. These consisted of the division of some rooms and the installation of a goods lift (later replaced by a small passenger lift).

At its first meeting the Council established The Physics Trust Fund under a trust deed. The object of this trust was to finance the educational and scientific work of the Institute and Society including the production and publication of its scientific journals and its scientific meetings. The Trust was recognised as a charity by the Commissioners of Inland Revenue and this had advantages as the parent body could not rank as a charity because it had a professional clause in its constitution.

In 1961 Council decided that the first anniversary of the amalgamation should be marked by a number of functions. A Presidential Address entitled 'The Development of Physics' was given by Sir John Cockcroft. A first Annual Dinner was held in the Park Lane Hotel in London, attended by 248 members and guests. The first Annual Representatives Meeting was held the following day, the object being to keep Council in touch with the members.

At the annual anniversary meeting in 1963, there was an innovation: the social occasion was described as 'the annual dinner, soirée and dance' and it is reported in the *Bulletin* that the dance in the Park Lane Hotel in London continued until 1 am. This was clearly considered a success as the same pattern was followed in 1964 and dancing again continued until 1 am. But in 1965 the dinner was transferred from the Park Lane Hotel to the Savoy Hotel and there was no dance. In that year, 'after dinner, members and their guests inspected exhibits of general and scientific interest', and there has been no mention of a dance in any subsequent year.

The annual Physics Exhibitions

The long established tradition of the Physical Society's exhibitions of scientific instruments and apparatus was maintained under the amalgamation. The 44th in the series was held in 1960 in the Royal Horticultural Society's halls in London; there were 140 exhibitors and an attendance of over 20,000. It was the policy of Council that these occasions should remain important scientific occasions and not become commercial displays. The President wrote personal letters to the chairmen and managing directors of exhibiting firms asking for their assistance in this, and he got an encouraging response which satisfied the Exhibitions Committee that there was general support for restricting the exhibitions in that way. The 1960 exhibition made a profit of 40% of its turnover of £20,000.

The next four annual exhibitions were again held in London and were as successful as ever. The stringent measures taken to limit commercial exploitation of the exhibitions in order to preserve them as scientific occasions continued to be warmly welcomed, but the Committee decided that these occasions should occasionally be held outside London and it was agreed that the exhibition in 1965 would be in Manchester. This move to do things occasionally outside London was reflected in the decision in 1962 to hold the anniversary meeting and the annual dinner in Harrogate.

The 49th annual exhibition was therefore held in the Manchester College of Science and Technology. Considerably less space was available than in London and refereeing had to be very strict, but it was subsequently decided that the experiment of moving the exhibition out of London was on the whole a success, though 'it would be inadvisable to repeat it more often than once in five years'.

The 50th exhibition in the series, held in 1966, was considered a special occasion and was moved from the traditional site in the Horticultural Society's halls to Alexandra Palace in North London. The move, dictated by increasing pressure on space, proved to be highly successful with nearly twice the area available for exhibits as well as improved ancillary services. Of major interest was the display of instruments from France arranged by the Centre National de la Recherche Scientifique. However, by 1967 the exhibition made a profit of only £5000 and concern for its future in the face of commercial competition was being expressed.

Professional matters

Whereas the former Physical Society was primarily concerned with the dissemination and discussion of physics through its meetings and publications, the Institute was formed so that it could consider professional matters as well. It was therefore logical that the new Council should decide to formulate a code of professional conduct and a committee was set up to make recommendations. Enquiries were being received about terms and conditions of service, consultancy fees, suitable salaries for various posts, the availability of special courses, the prospects for physicists at home and overseas, and part-time work for those who had retired early. However, it became clear that the task

Aerial photograph of CERN taken in 1967 (© CERN).

of formulating a suitable code was very difficult. In 1964 the first draft was produced by the committee and it was circulated to about 50 senior Fellows with experience in government service and industry. From the many comments received the committee stated that 'this project is one which may take some time yet to complete to the satisfaction of the majority of members' and decided to produce a second draft. Several years later, it was accepted that the code 'would still need much further time and thought', but it was agreed, as an interim measure, to compile notes for the guidance of members on various aspects of employment. Four documents were made available: (i) terms of engagement for full-time employment, (ii) consulting work, (iii) writing for the lay press, broadcasting and television, and (iv) working for commercial publishers. At the same time information about careers was regularly made available and careers conventions were attended throughout the country.

The tradition for both general and specifically professional conferences and lectures had always been strong in both the Physical Society and the Institute, and the tradition was maintained in the amalgamated body. However, it was noticeable that there was increasing international involvement. There was a major international conference on magnetism in Nottingham, an international meeting on spectroscopy in Exeter, and an international conference on physics education in London.

Named lectures such as the Guthrie and the Rutherford ceased to be lectures and became medals and prizes after 1965. Council also decided during the 1960s to make two further awards: the Glazebrook Medal and Prize to be awarded annually for conspicuous services in applying physics, and the Bragg Medal and Prize for distinguished contributions to the teaching of physics. The first was to be awarded in 1966 and the second in 1967.

Fulmer Research Institute

An important event of the mid-1960s was the acquisition of the Fulmer Research Institute. Founded in 1946 by Colonel Wallace Devereux, one of the most entrepreneurial of British metallurgists in the 20th century, FRI was a contract research company primarily intended to carry out work for British industry. As a result of the policy that all profits should be ploughed back into FRI, it grew at a steady, if unspectacular rate, until the annual turnover rose to £200,000 in 1961.

In 1964 FRI was put up for sale and it occurred to the Honorary Treasurer, Dr James Taylor, that FRI might be owned by the Institute of Physics and the Physical Society, which would then become the first learned society to have a research company doing research projects on a commercial industrial basis. An advisory group set up by Council to study the suggestion recommended acceptance of Dr Taylor's imaginative and bold proposal and FRI was purchased at a cost of £100,000, to be paid, interest free, at £10,000 a year over ten years from the Fulmer profits.

Edwin Liddiard, FRI's first Chief Executive, wrote in 1965:

> 'The Management and staff of the Fulmer Research Institute are fully aware that the acquisition by the Institute of Physics and The Physical Society presupposes that some revenue will be available from the Fulmer Research Institute to support

the work of the Institute and the Society. There seems every prospect that this will be possible without serious interference with the expansion of the Fulmer Research Institute. The Institute of Physics and the Physical Society can expect capital appreciation with the passage of years'.

FRI was administered by a Board of Directors which included the Executive Secretary and the Honorary Treasurer of the Institute of Physics, and *ex officio* the President of the Institute. There was an independent Chairman of the Board. The Director of Research was the senior Executive Director, initially Edwin Liddiard until his retirement in 1969 when Dr Eric Duckworth succeeded him. Under Dr Duckworth expansion and growth continued and by 1970 the annual turnover had risen to £360,000. This increase in profitability meant that the covenanted annual payments made by FRI to the Institute were such that the cost of FRI to the Institute was paid off by 1975 and the Institute had then acquired a valuable asset at no actual cost.

Despite turnover rising to £6 million by the end of the 1980s, it became clear that FRI required more continual investment than the Institute could provide. Council, under the Presidency of Dr Cyril Hilsum, realised that it was now less appropriate for a learned society such as the Institute of Physics to own its own research laboratories, and that it should divest itself of the complete business. This was accomplished in 1990.

A number of members suggested that physics education was a particular gap in the coverage by the journals. Council therefore decided to establish a journal of *Physics Education* and plans were made during the year 1964 to begin publication late in 1965. The journal would be primarily for sixth-form physics teachers, but it was hoped that it would also be of value to teachers in universities, technical colleges and training colleges.

On 18 November 1965, the Institute and Society sustained a severe loss in the sudden death of its Secretary, Dr H R Lang. He had been Secretary of the Institute of Physics until the time of the amalgamation with the Physical Society, and from that time had been the Secretary of the joint Institute and Society. He had expressed a wish that no formal obituary of him should be published in any of the Institute and

James Taylor (Honorary Treasurer, 1956–1966, and President, 1966–1968).

Society's journals, but Council put on record its great appreciation of his long and devoted service. He was a remarkable man who saw the Institute and Society through many changes and he provided a measure of stability throughout the years.

The year 1966 saw notable changes. A new President, a new Vice-President, an Honorary Treasurer and an Honorary Secretary all took office on 1 October. Although Sir James Taylor was new to the Presidency he had been Honorary Treasurer since the amalgamation and had been Honorary Treasurer of the Institute of Physics for some years prior to that. P T Menzies succeeded him as Honorary Treasurer.

Louis Cohen (Executive Secretary, 1966–1990).

Dr R Press became Honorary Secretary in succession to Dr C G Wynne, who had served since the amalgamation and had previously been Honorary Secretary of the Physical Society for some years. Dr J V Dunworth succeeded Professor Brian Flowers as Vice-President for Publications.

Norman Clarke had resigned as Deputy Secretary of the Institute during 1965 to become the first Secretary of the Institute of Mathematics and its Applications. On the death of Dr Lang in November 1965, Dr A C Stickland as Deputy Secretary had carried out the duties of Secretary, as well as being responsible for the publications, until Dr Louis Cohen was appointed as Secretary and took over on 1 July 1966. He was to serve until his retirement in 1990.

On 18 September 1967 Sir John Cockcroft died at the age of 70. He was one of the pioneers of nuclear research in Britain and was head of the Atomic Energy Research Establishment at Harwell in the immediate post-war years:

'He was born in Todmorden and went to Todmorden Secondary School where he showed all-round competence from mathematics to cricket. After active war service in France, he entered Metro-Vic as an apprentice and was sent to Manchester University to read electrical engineering. He showed such talent that he was persuaded to go to Cambridge, first as an undergraduate to read mathematics and subsequently to work in the Cavendish Laboratory. His impact on the Cavendish was partly because he brought technical skills which were not possessed by physicists of Rutherford's generation, nor by most of Rutherford's pupils. The combination of engineering training and considerable mathematical ability made him an invaluable addition to the Cavendish team, especially as with his relaxed, friendly, unassertive temperament he was prepared to help anyone with anything. In 1932 he produced one of the most dramatic discoveries of that dramatic age of physics when, with E T S Walton, he demonstrated that accelerated protons could split atomic nuclei, a discovery for which he and Walton were awarded the Nobel Prize for Physics in 1951. He was President of the Institute of Physics from 1954 to 1956 and the first President of the amalgamated Institute of Physics and the Physical Society from 1960–1962'.

The Royal Charter

The year 1970 was a memorable one in the constitutional development of the Institute and Society, the highlight of the year being the granting of a Royal Charter by Her Majesty the Queen on 6 November 1970. This was the culmination of three years' work on the Petition, the Charter and the Bylaws.

The Royal Charter, granted to the Institute of Physics by HM The Queen on 6 November 1970.

An Extraordinary General Meeting was held on 27 November 1968 to consider the proposal to apply for a Royal Charter and it was overwhelmingly approved. It became clear that the pathway to a Royal Charter is not one which can be traversed quickly. An extra General Meeting was held on 29 July 1969, after which the Petition was sent to the Privy Council. A Petititon from the Privy Council in favour of the Charter was presented to the Queen at Balmoral on 30 September and on 6 November 1970 the Royal Charter was granted.

This confirmed the position of the Institute and Society, in its new form as a chartered institute, as the pre-eminent body representing physics and physicists in the United Kingdom. There was another advantage: the Privy Council was not prepared to accept the cumbersome name 'the Institute of Physics and the Physical Society', to which many members had been objecting from the amalgamation onwards, and it became simply the Institute of Physics. The unsatisfactory abbreviation of IPPS became IOP.

During 1971 the final legal process of winding up the Institute and Society was completed. This change afforded a suitable opportunity for the Institute to receive a handsome Presidential Badge donated by Sir James Taylor, the former President and for many years the Honorary Treasurer—and also the benefactor who negotiated with ICI the interest-free loan which had enabled the Institute to purchase the Fulmer Research Institute. The Badge was worn for the first time at the annual dinner on 4 May 1971. It was also at that time that the derived symbol for IOP, often affectionately referred to as 'the bent paper-clip', came into widespread use on the stationery and other printed matter.

External relations

The years 1968–1969 were notable for two other events which were to have long-term effects. Relations with other professional bodies in science and engineering continued to develop. In December 1968 Council approved the formation of a Council of Science and Technology Institutes (CSTI) to form a link with other institutes and with the Council of Engineering Institutions (CEI). The CSTI was formally established in February 1969 with the Institute and Society given responsibility for the secretariat.

The Institute and Society became increasingly involved in the negotiations for the formation of a European Physical Society (EPS). A meeting of a steering committee for the Society was held in London in May 1967 with Sir James Taylor in the Chair. Progress was made on a draft provisional constitution and interim arrangements for a secretariat. It was recognised that the function of the Society should be that of a service organisation to provide an interchange of information, organise conferences and other meetings, and coordinate publications. It was in the following year, 1968, that the European Physical Society was formally established in Geneva with the Institute and Society as a foundation member. Dr Louis Cohen became its first Treasurer. The first Council of the European Physical Society was held during the inaugural science meeting held in Florence in April 1969 when the Institute and Society was represented by many members.

Membership

After the merger of the Institute and the Society, the membership increased from 9120 in 1961 to 11,273 in 1965. However, the income was very dependent on the number of members and after 1965 there were concerns about the financial position which were to become acute from the early 1970s onwards. The subscriptions had last been raised in 1964 and at that time overwhelming support had been given by members to the increase on the grounds that this was preferable to a reduction in activities. But the subscriptions had not been raised since then even though the detailed accounts from 1964 onwards showed the very narrow margins on which the Institute and Society operated.

The repeated references in Council to the need to increase membership led to the introduction of a new grade of membership, the Licentiate. This grade was intended mainly for assistants and senior members of laboratory technical staff who had followed recognised courses of study but below that required for the Graduateship of the Institute. This new grade became effective in 1965.

In 1971 the nation was suffering from serious inflation. The prospect of increased membership was not great and in fact there was a substantial decrease in both corporate and non-corporate members in that year. Because of inflation and the decrease in membership, 1971 saw the biggest deficit ever in the accounts, described in the Annual Report as 'a very alarming reversal in the financial position of the Institute'. To improve the position two attempts were made that year to increase members' subscriptions. Unfortunately, the necessary majority of 75% required by the Bylaws was not obtained from the membership on either occasion (in July 50% and in October only 26%). The following statement was therefore put to the membership:

'While certain economies might be made by a severe curtailment of activities of the Institute and services to members, the saving would be small compared with the services withdrawn. The solution to this financial problem is a considerable increase in subscriptions, and it is hoped that as a result of this explanation of the position when the next application is made members will respond by voting in favour of the motion, so that the finances may be restored to a sound basis'.

Financially the year 1972 was again difficult with very rapid inflation causing an increased expenditure without there being a corresponding increase in income. Prospects were improved, however, by the fact that in July 1972 members agreed by 81% to a 36% increase in their annual subscription to come into operation at the beginning of 1973. The deficit for 1972 was again substantial, but not quite as bad as the previous year. However, these deficits had meant that the reserves of the Institute had been seriously reduced at a time when it was desirable to build them up as the lease on the headquarters building in Belgrave Square was coming to an end in 1982.

In 1973 fewer conferences than usual were organised, but it happened that the average attendance improved as did the average number of papers. It was finally decided that the annual Physics Exhibition be discontinued beyond 1977. This was a sad end to what had been a long-standing tradition since 1905.

The outstanding feature of 1974 was the celebration of the centenary of the founding of the Physical Society of London. This was marked by three main events. First, a joint meeting in Jersey with the French Physical Society, which had celebrated its own centenary a few months earlier. Secondly, a centenary dinner at which the principal guest was the Secretary of State for Education and Science. Thirdly, on the day after the dinner, a programme of celebrity lectures followed by a special soirée. Other events marking the occasion were an exhibition at the Science Museum and the publication of a series of commemorative issues of *Physics Bulletin*.

There was a surplus in the accounts for the year 1974, but a deficit in 1975. Inflation was the major problem for the Institute, as for most organisations and for professional institutions in particular. Fortunately publications continued to prosper and to provide essential financial support for the Institute's other scientific activities. A connection of almost 50 years was finally broken when members of the publishing staff moved out of Lowther Gardens and all publishing activity, except the advertising department, moved to Bristol, occupying Netherton House in Marsh Street and parts of Techno House in Redcliffe Way.

Le Menu

*Les Roulades de Saumon Fumé Farcie à la Mousse
de Truite*

*La Tortue Claire des Iles en Tasse au Sherry
Les Paillettes Dorées au Chester*

*La Poitrine de Pintade Albufera
Les Croquettes Succès
Les Salsifis et Carottes Panachées*

*La Mousseline de Pêche Parfumée au Champagne
Les Lychees Rafraîchis
La Corbeilles de Friandises*

Le Café Savoy

Menu from the Centenary Dinner, Savoy Hotel, 7 May 1974.

The latter was eventually to become the headquarters of all publishing activity until 1997.

There was continuing reference to inflation in the Council reports and membership was down 2% in 1976. Fortunately publications flourished and were a source of income for the Institute, thanks to their worldwide sales.

There was new vitality in the Education Department with Professor Charles Taylor as Vice-President for Education.   Physics at Work exhibitions around the country presented real-life applications of physics to a significant number of school pupils and these continue to be one of the most consistently successful activities of the Institute. There was growing interest in the country in maintaining educational standards and the Department was heavily involved in the preparation of reports, in collecting statistics and in the organisation of committees and working parties, stimulated by the intervention of the Prime Minister in educational matters in his Ruskin College speech during the year.

1976 also saw the retirement of Dr Press as Honorary Secretary after completing ten years and he was succeeded by Professor Roland Dobbs. The following year saw the retirement of Dr H Rose after serving five years as Honorary Treasurer and he was succeeded by Professor J M A Lenihan.

As the lease of 47 Belgrave Square was due to end in 1982, the future accommodation for the headquarters was beginning to be much discussed and efforts were made to purchase a large, freehold property for use by the Institute.   One was found near Regent's Park, but the Institute was gazumped.   No other property was found, so the lease of 47 Belgrave Square from the Grosvenor Estates was extended at a favourable price in 1977 in order to provide the Institute with a headquarters building until 2050.   Plans for its restoration and improvement for the benefit of the membership were prepared.

Retrenchment

In the late 1970s two million were out of work.   The earlier rise in oil price was having its effect and a recession was affecting trade and industry in Britain and throughout Europe; it was reflected in Institute affairs. In 1977 both the Physics Exhibition and the Annual Dinner had ceased, but in 1978 there was another deficit in the Institute's accounts. This alarming reversal compared with 1977 was due to a considerably reduced surplus on publications from £208,313 in 1977 to £78,572 in 1978.

Despite the seriousness of the situation the Annual General Meeting failed by a narrow margin to approve increases in subscriptions that would have been necessary to maintain the Institute's services to members. Fortunately, the financial crisis which would have followed was averted at an Extraordinary General Meeting when small increases in members' subscriptions were approved. Since these did not cover the increased costs due to domestic inflation, staff redundancies became inevitable. The membership in the corporate grades showed a small increase, but this was less than the decrease in the non-corporate grades, so there was an overall decrease, a further reason for depression.

Retrenchment was essential. During 1979, the staff at headquarters had to be reduced by about 20%, resulting in reduced service to members in the organisation of meetings, conferences and exhibitions. The administration of the branches from headquarters was largely devolved to branch committees, who became responsible for their own mailings and accounts. Membership in the corporate grades fell by 60 to 9166, and in the non-corporate grades by 202 to 6076, giving a total of 15,242, a decrease of 1.7% from the previous year.

The following year, 1980, was one of consolidation in the face of the decline in activities and the reduction in staff at headquarters. Membership continued to fall, by about 3% in the year. On the positive side, a start was made on the refurbishing of the headquarters building, following a positive response from members to an appeal for funds for that purpose. There was, however, a surplus for the year of £180,000 which showed that benefit was resulting from the stringent economies and staff reduction, and this surplus rose to £459,000 in 1981 mainly due to the increased surplus from publishing activities.

A new, simplified structure for membership was introduced in January 1982. The change, as well as the reduced subscriptions for members under the age of 30, was designed to aid recruitment, particularly of new graduates, and arrest the continuing decline in membership. Sadly the new structure failed in its object and by the end of 1982 there was a further decline in the total membership of nearly 5%. The membership figures throughout these years fell from 16,012 in 1971 to 12,289 in 1984.

Despite the problems over membership and finances, the Council established a new standing committee for Industry in October 1982 and agreed to appoint an Industry Officer in 1983; there was a further extension of services to schools and an outstanding conference on

'The Neutron and its Applications' to mark the 50th anniversary of its discovery; the Barber Trust Fund provided valuable support to assist 40 young people to attend international conferences overseas.

The years of recovery

In the distant past, Major C E S Phillips served as Honorary Treasurer for 20 years from 1925–1945; Sir James Taylor served as Treasurer for 10 years from 1956–1965 before becoming President; P T Menzies and Dr H Rose both held the post for 6 years. It was noticeable that in the late 70s and early 80s, the Treasurers were surviving for very much shorter periods! It cannot have been an enjoyable time for them when the membership was continually falling; when the finances seemed always to be in difficulty; when inflation was hitting the country hard; when it was necessary to make staff redundant; when attempts to increase income by raising subscriptions seemed doomed to failure when proposed to the membership. By contrast, the mid-80s became years of recovery, admittedly slow recovery, but nevertheless the tide was beginning to turn. The membership began to grow again and small surpluses replaced the long series of deficits. It was noticeable that Treasurers began to last longer in office! That a change was coming was revealed in the membership figures: 12,289 in 1984 became 14,723 in 1990.

Refurbishment of the headquarters at 47 Belgrave Square, providing an additional committee room, better catering facilities and an enlarged dining room in the basement, was completed in 1984. The computerisation of membership records and accounts was completed. Physics at Work exhibitions were extended and a regular magazine for physics teachers and schools, *Snippets*, was published for the first time. Some financial assistance was given to the Association for Science Education to encourage science at the primary level. At the secondary level the Institute's report *Girls and Physics* was in heavy demand and received favourable comment. It was noticeable that attendance at meetings showed an increase in the number of participants from the industrial sector, roughly 30% of the overall total.

Professor E R Dobbs retired in 1984 as Honorary Secretary after serving eight years and was succeeded by Professor D H Martin. The Institute continued to be active in the affairs of the European Physical Society. One of IOP's representatives on EPS Council, Dr G H Stafford, served as EPS President from 1984–1986.

After a year without an increase in fees, members agreed to an increase in line with inflation, something they had refused to do ten years previously. An increase in the staffing of the Education Department allowed the Institute's activities in education and careers to be significantly broadened and intensified. An important event in 1984 was the inauguration of the Schools' Affiliation Scheme, which sought to improve contacts between physics teachers and other professional physicists through the medium of the Institute. For a relatively modest affiliation fee, school and college physics departments received multiple copies of *Physics Education* or *Physics in Technology*, careers booklets, pamphlets and wallcharts, the newsletter *Snippets* and other educational material either free or at cost price. The scheme was seen as being of particular benefit to younger and less experienced teachers and it met with a good response with over 600 schools joining in the first year. However, 1984 was the year in which the Institute's Graduateship Examination was set for the last time. Since its introduction in 1952, 3261 candidates had entered and 1203 had passed at honours degree level.

In January 1985 the Privy Council approved the use by corporate members of the term Chartered Physicist and the designatory letters CPhys could be used after their names. In July 1985 a formal application was made to the Engineering Council, through the Vice-President for Membership, Mr Arlie Bailey, for the Institute to become an 'Institution Affiliate' of the IEE, which was a nominated body to approve applicants for CEng, whereby suitably qualified physicists could be designated Chartered Engineers. In 1987 the negotiations came to fruition. 1988 saw the first IOP members registered as Chartered Engineers by the Engineering Council through the route now open to IOP members who have the appropriate professional experience.

The end of the eighties

The years 1988 and 1989 were successful ones; they were to be the last two full years for Dr Louis Cohen as Executive Secretary before his retirement due in 1990. The difficult years of the early 1980s were now passed; the finances were now in a satisfactory state, thanks mainly to the success of IOP Publishing under its new management. The membership was growing and it was significant that the pattern was changing—close to half of the members under 40 years of age were employed in industry. The number of groups covering areas of applied physics and engineering physics had been encouraged by the Industry Committee and this led to two new divisions, one for Engineering Physics and one for Applied Optics.

1988 saw the launch of a new members' magazine, *Physics World*, replacing *Physics Bulletin*, and it was immediately well received. It was produced by IOP Publishing in Bristol with a full-time staff of five led by the Editor, Philip Campbell, who used his broad editorial freedom to create a lively and responsive magazine for members, dealing with all aspects of physics and physics-based technology.

In 1989 a new Corporate Affiliates scheme was started which allowed physics-based companies and laboratories which have interest in the progress in research and development in physics and physics-based technologies, and in the future supply, quality and training of graduate physicists, to support and participate in the Institute's programmes. There were 16 companies affiliated before the end of the year and there was a successful two-day Corporate Affiliates Conference in November 1989 at which the education and training of physics undergraduates was examined in detail by representatives of affiliated companies and institutions of higher education.

Overshadowing all the activity in the Institute, there was sadness at the enforced absence through illness of the Executive Secretary, Dr Louis Cohen. It brought extra burdens on the staff, particularly to the Deputy Executive Secretary, Maurice Ebison, who became Acting Chief Executive.

Dr Cohen retired in 1990 after 24 years of distinguished administration of the Institute. He had steered the development of the Institute from the newly forged merger of the previous Institute of Physics with the Physical Society, through the granting of a Royal Charter, through some difficult years when the membership was declining and there were repeated financial crises, to a strong and growing position as both

Louis Cohen's retirement dinner at the Café Royal, 1990.

learned society and authoritative professional body. It was particularly sad that his illness had resulted in the amputation of his legs and that he was not able to return to the Institute for his final days as Executive Secretary. He died, aged 71, on 28 July 1997 and the following is extracted from the full obituary published in the Institute's Annual Review for 1997:

> 'The British community of physicists owes an enormous debt to Dr Louis Cohen, who was Executive Secretary of the Institute of Physics for 24 years. During this period, the Institute was seen by physicists as serving their needs and fighting their causes increasingly well. Institute membership rose by more than a half within six years of his appointment in 1966.

> When Cohen came to the Institute, he found that the merger of the Physical Society and the Institute of Physics, each with a long history and its own traditions, had led to some resentment. He worked hard during his first years to make the new body successful, and gradually his personality and perseverance won over almost all the members. By 1970, the Institute was in a position to seek and subsequently obtain a Royal Charter.

He had just taken up his post when a most embarrassing incident occurred. The Institute's Council, during the interregnum between Executive Secretaries, had invited the Duke of Edinburgh to become an Honorary Fellow. It was only after the Duke had accepted that it emerged that election to Honorary Fellowship required the agreement of the corporate members in a secret ballot. Unfortunately, this agreement was not obtained and it fell to Louis Cohen to carry out the unenviable task of informing the Palace of the changed situation. Consequently, he made sure that the Institute's bylaws were changed to place responsibility for election to Honorary Fellowship solely with Council.

Cohen's management and negotiating skills, honed during his early career in industry, came to the fore in the 1970s when the lease on the Institute's premises in Belgrave Square had only a decade or so to run. He negotiated successfully with the Grosvenor Estate for an extension of 75 years at most favourable terms.

He always emphasised the importance of education in physics at all levels, both as mental training and as a vocational education second to none. Under his leadership, the Institute established a reputation for supporting physics teachers and bringing an awareness of physics to the general public. During his tenure, the Institute became a world-leading physics publisher. He oversaw the integration of publishing activities in Bristol and the establishment of the publishing division as a limited company. In the years before his retirement, this publishing arm made an ever-greater contribution towards the promotion of other concerns.

He initiated many collaborative ventures. He played a leading role in the establishment of the European Physical Society. He served for many years on its Executive Committee and was Honorary Treasurer from 1968 to 1973. He was also instrumental in setting up the Council of Science and Technology Institutes, a forum for the presidents and chief executives of scientific institutes and societies.

He was elected a member of the Institute of Physics in 1949 and a Fellow in 1956. He retired as Executive Secretary in

1990, by which time he was confined to a wheelchair. He bore multiple disabilities with a grace and cheerfulness that earned the respect and admiration of all who knew him'.

Now came the task of finding a new Chief Executive. The post was advertised, and an Appointments Committee interviewed a shortlist of distinguished candidates from a large number of applicants. Council confirmed the appointment of Dr Alun Jones, who took office on 2 July 1990. With a broad background in physics, publishing and management, it was felt he had the enthusiasm and commitment to take the Institute forward to the next millennium.

CHAPTER 8

PUBLISHING

Institute of Physics Publishing at the end of the 1990s describes itself as one of the most innovative physics publishers in the world—editorially dynamic, leading in technological sophistication and widely respected as a learned society publisher. Yet 125 years ago the first publication of its ancestor, the Physical Society—the *Proceedings of the Physical Society*—was a mere reprint from the *Philosophical Magazine* published by the family publisher Taylor & Francis. So even then the 'commercial' sector of learned publishing had an influence on the communication of physics information; this commercialisation of publishing was to increase greatly during the years to follow, as we shall see.

How, then, did a learned society, principally concerned with giving its members an opportunity to communicate their subject to their local peer group, evolve into an international publisher whose services are used and publications purchased throughout the world? How, too, did that development fit with the way the Society itself and its successors branched out?

The Physical Society—early days

The early years of the Physical Society were devoted to the communication of physics at meetings, and publication was closely allied to that activity. Members showed demonstrations and delivered papers at meetings and these were published—initially in *Phil. Mag.* and reprinted in the *Proceedings of the Physical Society*. In due course the *Proceedings* became a journal rather than a proceedings, accepting submissions which were then read at meetings. The link between meetings and publication was not broken completely until the beginning of the 1940s, when the general meetings of the early days were complemented by meetings on more specialised topics.

The early concern with publication recurs repeatedly during the first decades of the society.   For example, in 1877 the membership deliberated over whether the society should 'publish exhaustive digests of the work that has been done in connection with special physical problems—such, for instance, as the determination of the velocity of sound or the mechanical equivalent of heat ...'. It is worth recording the conclusions reached by the assembled Fellows:

> 'The Committee, while fully recognising the great utility which work of the kind proposed would possess, hardly see how it could be efficiently carried out otherwise than as an individual literary effort, which the society would not have any special power for furthering'.

In 1999 this would have been seen as an opportunity to open a new line of publications! We shall find other instances of this early reluctance to engage in adventurous publishing as we trace the history of the Society's and its successors' publications.

Quite early it was suggested that there was an important amount of published physics not read by English-speaking practitioners because of the language barrier—a barrier which the increasing use of English as the *lingua franca* of science has now all but abolished. Fellows of the young Society felt that this material should be made accessible to those not blessed with a knowledge of German, for example, and the eventual outcome was the publication of the Society's *Physical Memoirs*, translations of papers originally published outside the country. Volume 1 in 1891 contained translations of three papers by Helmholtz and one by van der Waals.

The *Physical Memoirs* did not have a long life. More can be said of the *Abstracts of Physical Papers from Foreign Sources*, whose first volume was published in 1895. Like the *Memoirs*, the *Abstracts* had the aim of making material first published abroad accessible to physicists in England, this time in summary form. Published by the same firm of Taylor & Francis, the *Abstracts* lasted through to the third volume in 1897, to be replaced in the following year by the

CONTENTS.

VOL. I.

— · —

Contents page of volume 1 of the 'Proceedings of the Physical Society of London'.

first volume of *Science Abstracts: Physics and Electrical Engineering*. Issued 'under the direction of the Institution of Electrical Engineers (IEE) and the Physical Society', it was published by the London firm of E & F N Spon Ltd. Keeping up with the growing body of published research literature was an issue even then, and the two partners had clearly identified a need, for over the next century the enterprise grew into the dominant secondary information service in physics (and still electrical engineering), eclipsing others such as *Bulletin Signaletique*, *Physikalische Berichte* and *Referativnyi Zhurnal* in its international impact. What is puzzling is that the service now known as INSPEC (Information Service in Physics, Electrical Engineering and Computing) is still provided by the IEE but without input from or control by the Institute of Physics. The Physical Society and later the combined Institute of Physics and the Physical Society contributed financially to *Science Abstracts* to the tune of £2200 in 1964, the last record in the accounts. The Institute was well represented on the *Science Abstracts* Editorial Advisory Panel, had a seat on the management board and made valuable contributions to the development of a new physics and astronomy classification scheme, but that is where its interest ended. Any ownership it once had in this vast intellectual property has long since ceased.

There are other such examples of publishing opportunities which presented themselves to the officers of the Society or its successors from time to time but which were not pursued with the vigour that they would be today, when publishing is considered a core activity of the Institute rather than ancillary to its main purpose of serving the needs of the subject and its members.

The Institute of Physics

The foundation of the Institute in 1919 was to meet those needs of physicists working in industry and government, as well as in academic research, which were not satisfied by the Physical Society. It was not long before here, too, publication was used as a tool for providing services to members. In 1923 the Institute launched the *Journal of Scientific Instruments*, 'dealing with methods of measurement and the theory, construction and use of instruments as an aid to research in all branches of science and engineering'. The first issue was introduced by J J Thomson, then President of the Institute. Interestingly it was a launch issue, with the Institute reserving a decision on whether to publish, subject to the results of what would now be called test marketing. The princely sum of 30s 0d (£1.50) was proposed for the annual subscription and Thomson suggested that a circulation of 3000 would break even. The circulation was eventually to go well beyond 5000 and the journal survives and prospers under its current title *Measurement Science and Technology*.

There is no one unique publishing model and here we might note that the idea for the journal was promoted by Sir Richard Glazebrook, the first President, when Director of the National Physical Laboratory (NPL). The journal was initially edited at the NPL under the supervision of a scientific advisory committee appointed by the Institute, and managed by a finance committee of the Institute with representatives of the NPL and the Department of Scientific and Industrial Research. The need for interdisciplinarity was recognised even then, with the desire to co-opt biologists, engineers, chemists and instrument makers, 'as well as physicists', on to the scientific advisory committee.

In publishing terms, the Institute proved more adventurous from the beginning than the Physical Society. The first Institute of Physics lectures on 'Physics in Industry' were delivered in 1923 and published under the 'Oxford Technical Publications' imprint by Henry Frowde and Hodder & Stoughton. The series ran for many decades and was later published for the Institute by Chapman & Hall. It ran out of ideas, so it seems, in the 1950s and was formally abandoned in 1963.

Another of these forays into non-journal publishing were the *Laboratory and Workshop Notes*. Published from the very first issue in the *Journal of Scientific Instruments*, they were clearly perceived to be of more permanent value. In 1949 the first volume of *Laboratory and Workshop Notes* was published for the Institute by Edward Arnold as reprints from the *Journal of Scientific Instruments*. Edited by Ruth Lang, they were to be issued thereafter on a three-yearly basis and lasted until 1967 when the publishers concluded that sales were too low.

One further example of publishing innovation from the Institute was the monographs for students. Launched in 1955, the series was originally intended for Higher National Certificate students but the target audience became students in universities and colleges, and also graduates 'to refresh or extend their knowledge'. The titles were diverse and very practical, including for example: *Soft Magnetic Materials Used in Industry* by A E De Barr; *Mechanical Design of Laboratory Apparatus* by H J J Braddick; and *Subjective Limitations on Physical Measurement* by C A Padgham.

One of the most popular was *Errors of Observation and their Treatment* by J Topping, which went through three editions and a further five reprints. The series was published for the Institute by Chapman & Hall. Despite what appeared to be a successful career it ran out of steam and was discontinued in 1965. The decision to close down the endeavour, and also the Physics in Industry series, was taken by the Council of the Institute and Society in 1963 as 'the need was now largely met by commercial publishers', promising however that the Books Editorial Board would 'continue to note any gaps which do not appear to be filled by commercial publishers, and advise commercial publishers of their existence'. Altruism indeed!

Having ventured into journal publishing with *Journal of Scientific Instruments* (*JSI*), the Institute was to take one further such step with the foundation of the *British Journal of Applied Physics*. The first issue was in January 1950, edited by H R Lang, and was welcomed by the President of the Institute, F C Toy, as a venue for announcing new applications of physics as well as developments of those applications already known. It was sent on its way with a grant from the Royal Society and the 'sympathetic support and encouragement'—but no money—from the Council of the Physical Society.

Intriguingly, it was an early example of a phenomenon of later decades known as 'twigging'. *JSI* had for some years been publishing occasional articles on applied physics, to the extent that for a time the title was enlarged to *Journal of Scientific Instruments and Physics in Industry*. 'Fission—to use a current term—has now taken place' said Toy, and the physics in industry part of *JSI* became the new venture. The Institute was to use the same technique of twigging several times later on.

Let us return briefly to the publishing activities of the Physical Society before we finish this account of the first 90 years or so of publishing.

The Optical Society was formed in 1899 and would have celebrated its 100th anniversary in 1999 had it survived. However, in 1932 it merged with the Physical Society and the interests of the optical community are now catered for by the Optical Group and Applied Optics Division of the Institute. A victim of the merger was the *Transactions of the Optical Society*, which survived for 33 volumes from 1899 to 1932, before being incorporated into the *Proceedings of the Physical Society* in 1933. The only visible impact on the latter was an increase in the number of issues from five to six a year. Well might one speculate what would have been the optical publishing scene in the 1990s had the *Transactions* survived. In the early 1970s the Institute came to an agreement with its sometime partner Taylor & Francis to merge the optics content of one of the Institute's core journals with *Optica Acta*, later to become *Journal of Modern Optics*. The Institute no longer has an interest in this, neither editorial nor financial. The strongest optics journals are now published by the Optical Society of America, and the strongest British journal by Taylor & Francis, and it is only during the last few years that the Institute itself has been able to claw back some of the losses in publishing coverage resulting from these earlier decisions.

More positive was the decision of the Physical Society to publish, from 1934, *Reports on Progress in Physics* as a 'series of reports on the state of physical science, which will include surveys of the progress in optical and allied matters'. In those days it was still possible to survey progress in the whole of physics, from sound to atomic physics, during one year in a single volume. Interestingly its main competitor then and now, the American Physical Society's *Reviews of Modern Physics*, chose to adopt a more specialised and selective approach when it was launched in 1929. The journal *Reports on Progress in Physics* prospered and grew in size, moving from an annual volume through several metamorphoses to bimonthly publication in 1969 and finally to its present monthly journal form in 1970.

Amalgamation and rationalisation

There is a natural divide in the Institute's publishing history, though perhaps it is more like a gradual transition than a sudden revelation of new purpose. The last new journal launched by the Institute or the Society—before the two bodies amalgamated—appeared in 1950. The

1950s themselves saw active book publishing with commercial partners, but this was beginning to fade out by 1960. When the Institute and the Society merged in 1960 to form IPPS (commonly so called because the full name was so unwieldy) the combined body owned four journals and a history—a history of adventurous experimentation but also largely of lost opportunities. In the mean time Robert Maxwell had come back from the wars and, after dabbling with Springer, had launched himself as Pergamon Press in 1951, to set the publishing scene for the next few decades. Twigging, journal specialisation and designer publishing became the mode from which many publishers worldwide, commercial as well as not-for-profit, prospered.

At this stage of the history, the journals were: *Journal of Scientific Instruments* (to become *Measurement Science and Technology*); *British Journal of Applied Physics*; *Proceedings of the Physical Society*; and *Reports on Progress in Physics*.

Publishing developed only slowly after the Institute and the Society joined forces in 1960 other than the integration of the editorial organisations of the two bodies under the supervision of Dr A C Stickland until her retirement in 1967, by which time the foundations for the next 20 years had been laid.

A number of Institute concerns were beginning to be reflected in changes in publishing. Being a professional body, the Institute had perforce to concern itself with professional standards and the way they were put across in secondary and tertiary education. The publishing product of this concern was *Physics Education*, decided upon in 1964 and launched under the editorship of Kevin Keohane in 1966. Its aim was to serve teachers of physics—across the sixth-form/first-year-university interface—and it has remained true to that purpose, though successive editors have tended to shade that focus 'up' or 'down' depending on their own views of the needs of the time.

A second concern—not unique to the UK and felt particularly in the USA—was that primary literature was overwhelming physicists' ability to keep up to date. A discussion of the place of reviews in the literature of physics was held within IPPS in the mid-1960s, and was the stimulus to the expansion and increasing frequency of *Reports on Progress in Physics* referred to earlier. It also identified the need to encourage technology transfer through publication, and led to a programme of invited applied physics reviews. Originally these were published in the successor of *British Journal of Applied Physics*, then as the *Review of Physics in Technology* and finally as *Physics in Technology* which, alas, quietly submerged its identity in *Physics World* in 1988 due to lack of support from individual and library subscribers.

More memorable for the history of the Institute's publishing was the decision to start the *Journals of Physics*. By the mid 1960s *the Proceedings of the Physical Society* had grown in size, the quality of the applied journals, *BJAP* and *JSI*, had been deliberately improved and a more rational division of material between the three journals had been devised by 1963. The imaginative stroke, executed during the Presidency of Sir James Taylor and with John Dunworth of the National Physical Laboratory as Vice-President for Publications, was to merge all of them into the *Journals of Physics*, parts A to E, from 1968. The first three parts were derived from the *Proceedings of the Physical Society*, which had split into three just before, and the last two from *BJAP* and *JSI*. The move was of fundamental significance, for it allied the Institute with the tendency towards specialisation in publishing, as we shall see later, rather than the more monolithic approach favoured by the American Physical Society with *Physical Review*. It also allowed the Institute to extend its influence in applied areas of publishing which the more purist approach of *Physical Review* or the *Proceedings of the Physical Society* would not have countenanced.

Publishing management

Equally significant was a change in the approach to managing publications. In the early 1960s, publishing was represented by an editorial office in 1 Lowther Gardens, in Kensington, London, a charming Victorian building rented from the Royal Commission for the Exhibition of 1851, with a resident ghost on the top floor—as sworn by many members of staff at the time. The leader of the small team there—more a family—was Dr Stickland, as she was known to everyone; no first names then. What are now the publishing, editorial and production functions were in the hands of the same person and the art studio was a tracing office in the attic. Marketing and distribution were done by the Institute's accounts department in 47 Belgrave Square.

As an indicator of the change to come the Institute appointed Cecil Pedersen as 'Director of Publishing' on the retirement of Dr Stickland in 1967. Publishing had arrived as something that the Institute did as part of its mission, but was also as a force with its own identity. It is worth tracking the history of this change in management approach for a couple of decades—for in societies these matters evolve slowly—before returning to publishing proper.

In 1967 the key determinants of policy were the Editorial Boards of the journals, and the Publications Committee reporting to Council as guardian of the Institute's interests. All the Honorary Editors were represented on the latter. Publishing was relatively simple then: marketing was minimal, office systems were paper based and the revolution of high-speed photocopiers, computers, electronic mail and digital text were undreamed of. Publishing was a relatively small affair, with some 15 or so people at 1 Lowther Gardens, exhibiting a dedicated but nevertheless fairly relaxed approach to working life. 'Staff' were always such and took no part in decisions: at meetings of the Editorial Boards, for example, they were always 'in attendance'. The decision to appoint a Director of Publishing lit a long fuse in two ways. Firstly it imbued that post with a potential authority never held by the Editor, and secondly it prepared the ground for more substantive changes when physics publishing grew, staff numbers swelled, journals proliferated and it was no longer possible to manage the Institute's publications in the traditional way.

As the number of journals increased and as the net revenue flowing into the Institute became a more significant part of its overall income, more attention needed to be devoted to the 'business' from day to day, something which a publications committee meeting a few times a year was not set up to deal with. A Publications Executive was set up in the 1970s, during the Vice-Presidency of R S Pease, composed of Honorary Officers of the Institute and senior publishing staff, but the decision-making route remained through Publications Committee to Council. By this time there was also a books committee, and matters became even more complicated when the Institute purchased the commercial book publisher Adam Hilger Ltd, with its own Board of Directors. After several more twists and turns, the limited company IOP Publishing Ltd was finally set up in 1986 as the sole vehicle through which the Institute's publishing was carried out. The Board of the company was chaired by a Vice-President of the Institute—then Professor Frank Read—and included members of Council and several executive directors including the managing director Anthony Pearce. The Board was still advised, on scientific matters, by two committees carried over from the previous regime: the Journals and Books Editorial Advisory Committees. Except for the disappearance of the latter two committees, this structure of publishing survives with only marginal changes in the composition of the Board.

These cumulative changes left publishing more responsive and more tightly managed, commensurate with the needs of the competitive 1980s and 1990s. It was also recognised that for effective management of what was now a business, there was a need for a more arms-length relationship between the Institute and what had become its biggest activity.

Early journal development

Such was not yet the case in the early 1970s, when the *Journals of Physics* had established themselves, but other things were happening, principally because of the strong growth in the amount of material that there was to be published. Commercial publishers such as Pergamon Press saw this as a huge opportunity to profit, and although the Institute continued to look on publishing as a service activity, it could not escape the logic. It had good journals and people wanted to publish in them. Submissions grew, as did pages published—from 6200 in 1967 to 7500 in 1969, for example. In 1971 we see the second example of twigging: *Journal of Physics C: Solid State Physics* spawned *Journal of Physics F: Metal Physics*, edited by R H Bullough. Sam Edwards was Chairman of the Publications Committee at the time and oversaw not only this particular process but an effort by the Institute to ally itself with the metals community through editorial collaboration with the Institution of Metallurgists. It was a worthy attempt to provide publishing services to another community, though it ultimately failed and in 1989 *Journals of Physics C* and *F* were merged again to form the present *Journal of Physics: Condensed Matter*.

More lasting was another twigging in 1975 when Denis Hamilton became honorary editor of *Journal of Physics G: Nuclear Physics*, grown out of *Journal of Physics A: Mathematical, Nuclear and General*. The journal survives with a slightly expanded title, witness to the Institute's stake in an area of physics publishing dominated by the American Physical Society and North-Holland/Elsevier. *Journal of Physics A* was to be the parent of other new journals later on, with *Classical and Quantum Gravity* in 1984, *Inverse Problems* (1985), *Nonlinearity* (1988) and *Network: Computation in Neural Systems* (1990) all laying at least partial claim to this parentage.

Collaborations

In parallel with this organic development of core journal publishing ran a number of collaborative ventures which added new facets to the showcase of journals published by the Institute. The first was *Physics in Medicine and Biology*, purchased from Taylor & Francis in 1972 and published with the Hospital Physicists' Association, as it was then called. A sister journal was launched in 1980 as *Clinical Physics and Physiological Measurement*, and a third medical physics journal was added in 1988 when the Institute took over the publication of the *Journal of Radiological Protection* for the Society for Radiological Protection. Still in the 1970s, following the build-up of plasma physics papers in *Journal of Physics D: Applied Physics* (*BJAP* as was), the Institute bought a half-share *in Plasma Physics* from Pergamon Press in an attempt to rationalise the publishing of plasma physics papers in the UK. Because of Robert Maxwell's somewhat chequered reputation, even then, the agreement had a clause allowing the Institute first refusal on the other half, should Maxwell cease to be chairman of Pergamon Press. This duly happened when he fell overboard off the Canary Islands, and the Institute purchased the remainder of what is now *Plasma Physics and Controlled Fusion* in 1992.

Other significant collaborations included the launch of *Nonlinearity* with the London Mathematical Society in 1988 to join forces rather than have the two societies launch competing journals with strongly overlapping scopes, working with the European Optical Society over what are now *Journal of Optics A: Pure and Applied Optics* and *Journal of Optics B: Quantum and Semiclassical Optics*, and with the Institution of Electrical Engineers on *Distributed Systems Engineering*. Somewhat off the beaten track for the Institute, but nevertheless not too far

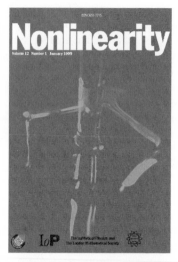

A selection of the Institute's journals.

removed from its core objectives, is *Public Understanding of Science*, launched with the Science Museum in 1992.

A particularly fruitful association, though it led to only two new journals, was with the European Physical Society (EPS). The EPS, ever since its foundation in 1969, had as one of its aims to make a difference to the way physics is published in Europe, seeking to introduce common approaches and standards and, eventually, fostering a European counterbalance to what even then was the dominance of the American *Physical Review*. The initial approach was to introduce a hallmark, the Europhysics Journal label, which certified that the holder satisfied certain criteria concerning refereeing standards, international composition of the editorial boards of journals, no page charges and the active contribution of boards to editorial policy. On most counts the Institute's journals were already in compliance with the criteria. However, the editorial boards were almost wholly British, and the majority of authors and referees were from the UK.

The 'Europhysics Journal' label, which all the *Journals of Physics* had acquired by 1973, led to the opening out of the editorial boards, with the introduction first of board members from Continental Europe and then from other parts of the world where physics research was thriving, and eventually the appointment of honorary editors from overseas also. As a consequence, the proportion of overseas referees and authors also increased greatly and is now mainly from outside the UK. The influence of the EPS on the shape of the Institute's journals, directly or indirectly, has therefore been profound.

When the EPS was formed, the Institute's bid to publish its membership publication, *Europhysics News*, was not successful. There are, however, two journals on which the Institute collaborates with the EPS. The first, *European Journal of Physics*, was launched with George Series as honorary editor and did reflect the broader publication aims of the EPS. It was launched as a complement to the *American Journal of Physics*, for scholars and teachers of physics in tertiary education, but also as the cornerstone of a European physics journal enterprise to rival the Americans'. Alas, it did not grow into a larger edifice of European research journals, though the European vision is still alive as exemplified by the merger of *Zeitschrift für Physik* and *Journal de Physique* in 1998 to form the *European Physical Journal*.

An earlier merger of publications under the EPS umbrella had been under discussion for several years in the early 1980s. The scheme was to create a European counterpart to *Physical Review Letters* by combining the letters sections of the *Journals of Physics* with the existing letters journals of the French and Italian Physical Societies. All seemed to go well until the final hurdle, when the proposals were presented to the Institute's Publications Committee by Roy Pike, then Vice-President. To a man (and honorary editors of Institute journals have always been men!) the assembled editors of the *Journals of Physics* resisted his urging and insisted that the place of letters was with the full papers in their subject-specific journals. A compromise had to be found and the Institute ended up contributing cash and marketing, while the French and Italian partners merged their letters journals into *Europhysics Letters*. The fourth partner was the EPS which was, and continues to be, responsible for the scientific direction of the journal. During the formative years of the journal other European societies contributed funds, and the journal stands as a monument of what is possible, being modestly successful though some distance from the original objective of making *Physical Review Letters* feel insecure. Would a different Institute decision have made a difference to the project? Who knows, but the fact that the Institute felt unable to contribute a material part of its publishing—the letters sections—to the new venture continued to colour continental views about the Institute's commitment to European collaboration for years to come.

Book publishing

The Institute's journals have an unbroken history of 125 years. The same cannot be said of the two other current mainstream publishing activities, books and magazines. Let us start with books. We have seen that for the first 75 years or so not very much happened in book publishing that has survived until today. Leaving commemorative volumes and conference proceedings to one side, there were only the innovative adventures of the Institute into student monographs and 'recycling' of material from *Journal of Scientific Instruments*, and these did not survive the 1960s, having effectively been handed over to commercial publishers. In the 1960s and early 1970s, after the amalgamation, spirits seemed to be a little more adventurous, and two signposts are worth pointing out. The first is the publication in 1966 of *Physical Basis of Yield and Fracture*, hardly the stuff that publishing revolutions are made of but in fact the first title in the *Institute of Physics Conference Series* whose volumes have now exceeded 160. This was the start of a major enterprise of publishing the proceedings of conferences—both within the series and outside. At first these were conferences organised by the Institute, but later other major and minor international conferences were tackled culminating in such epics as the European particle accelerator conferences published in 1996 and 1998 as weighty paper tomes with accompanying CD-ROM versions for conference delegates. The Institute did not go to the same lengths as World Scientific or some other commercial publishers, however, which built up whole publishing programmes on the back of conference proceedings. Instead it adopted a more flexible if less consistent approach, which included attempts such as the short-lived series of *Conference Digests* of the 1970s, the *IOP Short Meetings* of the 1980s and the continuing publication of proceedings in journals where appropriate. One of the main criteria for making such decisions was how best to serve the needs of particular communities, for example the groups and divisions of the Institute in the case of *Short Meetings*.

A second trend was signalled in 1972 when a fairly unkempt collection of quotations, essays, cartoons and other previously published material arrived at the publishing offices for consideration. The apocryphal story is that the late Robert L Weber had submitted the typescript to 47 different publishers before coming to the Institute. Intrigued, the publishing staff devoted some considerable time to sifting, sorting and selecting before eventually agreeing to take a publishing risk. A brainstorming session produced the happy title *A Random Walk in Science*, and after publication in 1973 the book proceeded to sell well over 40,000 copies direct and through book clubs, not to mention translations. Robert Weber turned out to be one of the Institute's more prolific authors, producing two sequels to *Random Walk* and two further titles. He had started a trend, and *Authenticity in Art* by Stuart Fleming in 1975 and *Harvest of a Quiet Eye* by Alan Mackay in 1976 were early successors as books of more general interest. Originally billed as Institute of Physics books for members, they turned into precursors of a list of general science books. Book purchases by Institute members were not enough to underpin the Institute's book publishing, but there was an appetite amongst the wider, mostly scientifically literate, public for general-interest, popular and even quirky titles on scientific topics, and authors have come to the Institute because of the early reputation it had acquired in this niche. By now these have formed a list in its own right, ranging from the books of Euean Squires on basic physics, through the stimulating *Riddles in Your Teacup* by Home and Ghose to the science politics of *The Quark Machines* by Gordon Fraser.

Apart from these early developments book publishing was fairly haphazard until 1976 when R S Pease, as Vice-President for Publishing, modestly declared in the Institute's annual report that 'During the year the Institute purchased Adam Hilger Ltd, a small scientific, technical and medical publishing house, which in the past has specialised in applied science titles'. An understatement if ever there was one, for it was only the second time that the Institute had engaged in corporate acquisition, the adoption of the Fulmer Research Institute (see chapter 7) in 1965 having been the first.

Adam Hilger Ltd was an instrument maker who started to publish manuals and then books in optics. In time, its book list acquired its own identity, and by the time of the purchase from Rank Precision Industries it had diverse interests ranging from spectroscopy through pure mathematics to astronomy and amateur geology. The list reflected the interests of Adam Hilger's managing director, Neville Goodman, who was a keen walker and amateur astronomer.

The opportunity to turn the Institute into a book publisher with a diverse list was one of the main reasons put to Council for making the acquisition—it offered some protection against what was even then felt to be the threat of decline in revenue from journals.

For a number of reasons book publishing has not served the Institute as well as had been hoped. The experience of publishing journals was not easily transferable to books, whose marketing, sales and distribution channels are quite different. As a result, book publishing was buffeted more than journals by changes of editorial direction, organisational approach, and sales and marketing strategies. Initially, development of a books list was not well targeted and included the publication of translated titles for which there was limited demand. Subject areas expanded and contracted, as did the number of titles, not always of top quality. The results of these changes were reflected in the financial performance. When the accounts were examined in detail, losses were seen to be greater than expected and the Institute had to decide whether it wanted to publish books at all. A Books Working Party in 1993 led by the President, Clive Foxell, recommended that it should continue to do so, but at a much more modest level. This safeguarded the immediate future of books within the Institute but it was to take another three years before the Institute agreed to a more confident and expansionist approach to books.

The above has simplified the books story and has overlooked one major perturbation. In 1984, following some market intelligence acquired in the Llandoger Trow public house in Bristol, the Institute, under the Vice-Presidency of Roy Pike, seized an opportunity to purchase the ailing John Wright group of companies. This was a medical book and journal publisher, an established printer who had printed Institute journals in previous decades, and a distributor. The purchase initiated three years of, alternatively, excitement and trauma for both publishing and the Institute. Finding itself suddenly a medical publisher, the Institute had much to learn about this market. The Adam Hilger books operation had to be integrated with John Wright books in commissioning, sales and marketing, as did the John Wright journals with the Institute's own journals in the editorial and production functions.

The original recommendation to Council—to dispose immediately of the printing company—was not followed through. The printing company needed much investment to bring its technological infrastructure up to date, resources which the Institute did not have. Sadly, the company went into voluntary liquidation in 1985. A management casualty of the change was Cecil Pedersen, who resigned in September 1984, to be succeeded by Anthony Pearce in February 1985. To manage these newly complex publishing arrangements better, the vehicle of Adam Hilger Ltd was transformed into IOP Publishing Ltd in 1986 under the chairmanship of Frank Read, to manage all of the Institute's publishing on behalf of the Council of the Institute.

The end of the episode was described by Frank Read in the 1987 Annual Report of the Institute: 'A key event during the year was Council's decision to dispose of the John Wright medical imprint to Butterworth Scientific ... to concentrate Institute resources on expanding its physics publishing'. Put differently, to make the medical publishing side profitable would have required investment at a level which would not have left sufficient resources for the Institute's core purpose. The

A selection from the Institute's books list.

Institute managed to cover most of its costs, including those arising out of its responsibility for securing an empty building formerly occupied by the print and distribution parts of the group. It disposed of its interest in distribution to a management buy-out a little later, and found itself back with its core business of physics journals and books.

The Philadelphia office

The books story is rounded off by two further decisions. The first was to open an editorial office in Philadelphia, USA, in 1990. Initially it was a one-person operation with Sean Pidgeon as commissioning editor. Soon, with changes in the North American marketing and distribution arrangements for books and journals, the office took on responsibility for North American sales and marketing to reflect properly the importance of this market to the Institute. The second was the decision to embark on a programme of major reference works. The

'Twentieth Century Physics', co-published in 1995 with the American Institute of Physics.

first title was launched in 1994—*Neutrons, Nuclei and Matter*. Others followed, including the three-volume *Twentieth Century Physics* and the industrially targeted *Handbook of Surface Metrology*.

Physics World

We need now to retrace our steps to uncover the origins of the third strand of the Institute's current publishing activity. Back in 1950 the Institute started publishing a small eight-page A5 newsletter of notes and notices, giving Institute announcements and news plus occasional aphorisms collected by anonymous editors. At the time of the amalgamation of the Institute and the Physical Society in 1960 this had grown to a small membership magazine, the *Bulletin of the Institute of Physics and the Physical Society* (no wonder a new name was needed!). This now carried feature articles and general science news as well as house news and information. Many articles dealt with educational topics until 1966 when *Physics Education* was launched as the Institute's education magazine.

The rethinking that led to the formation of the *Journals of Physics* in 1968 also brought the recognition that a modest society bulletin was less than the members of the Institute deserved. In 1968 it was relaunched as *Physics Bulletin*, a physics magazine primarily for members, but also one which for the first time was available on public subscription and could be seen as the Institute's window on to the world of physics. To defray the more substantial production costs, advertisement space was sold more vigorously by the recently appointed advertisement manager, Stephen Sadler, to the extent that the then President, M R Gavin, gave a dinner for the magazine's staff when the advertisement revenue for a single issue first exceeded £2000—double the revenue for the whole of 1967 but now less than the cost to advertisers of a single page in *Physics World*, the successor to *Physics Bulletin*. The magazine had a happy time for most of the next 20 years but was never likely to make a serious dent in the reputation of its prime rival, the American *Physics Today*. Lack of resources was one reason for this; indeed, in 1980 the magazine was thinned out and the number of issues reduced because of the financial difficulties of the Institute at the time. A second reason was that this was meant to be a magazine, yet it was published by an Institute whose strength was learned research journals; their respective cultures are very different but to a degree the journals approach showed through in the magazine. It did have two stable mates—*Physics Education* and *Physics in Technology*—but neither of these were 'real' magazines and now *Physics Education*, the only survivor of the two, is seen more as a journal than a magazine.

The last reason was that *Physics Bulletin* never had an editor in more than name: those who held the title, Cecil Pedersen and Kurt Paulus, had many other responsibilities, and the 'staff editors', Frances Fawkes, Neil Warnock-Smith and Jill Membrey, were never given the status or responsibility that an editor of a leading physics magazine would aspire to.

Physics World.

These three disadvantages were removed in October 1988 when *Physics World* was launched, replacing both *Physics Bulletin* and *Physics in Technology*. This time the job was done properly, with market research on the likely advertisement potential, with editorial and design consultancy and with the appointment of Philip Campbell as full-time editor with full authority over the content of the magazine. The move had the full support of the Institute, though perhaps mixed with a little trepidation, and there were occasions when the Council might have preferred a closer rein over the editor—but it held its tongue! Philip Campbell eventually took a further step in his career when he returned to his former publication *Nature* as editor in succession to John Maddox. Under Philip Campbell and his successor Peter Rodgers *Physics World* has grown greatly in reputation and stature.

Magazines

As we shall see later, the late 1980s were years of publishing innovation and the same spirit was to drive the 1990s. The launch of *Physics World* turned into a springboard for the Institute's diversification into the heady world of advertisement-driven technical magazines. *Opto and Laser Europe* was launched in 1992. *Scientific Computing World* followed in 1994, *Fibre Systems Europe* in 1996 and *Vacuum Solutions* in 1997. Such was the reputation that the Institute acquired in this publishing niche that when the Royal Astronomical Society decided to relaunch its own member publication as *Astronomy and Geophysics*, the Institute was the chosen publisher, and *CERN Courier* followed in 1998.

This adventure into markets traditionally dominated by commercial publishers was accompanied by a good deal of nail-biting on the part of the board of the Institute's publishing company. It was a strategy requiring much investment in an untried area, and one accompanied with a good deal of risk and the fear of reprisals from commercial competitors. Nevertheless, stoutness of soul prevailed and during the last few years magazine publishing has developed to deliver a service to the Institute's industrial constituency which its traditional publications did not convey.

There is a footnote worth recording, if only to make the point that the path to success has many diversions. One innovation was the suggestion the Institute should develop a series of technical newsletters, highly priced and with essential content for its readership. Both *Object Manager* and *Environmental Sensors* were launched in 1993 but ceased

A selection of magazines published by Institute of Physics Publishing.

publication in 1996 when it was recognised that they held no promise of success. *Noise and Vibration Worldwide* was purchased, also in 1993, with the intention of turning it into a newsletter and was sold again in 1996 when its commercial viability was seen to be too insecure. The episode illustrates that the Institute has, in the 1980s and 1990s, become much more adventurous in its publishing than one might expect from its venerable history. Inevitably, and without loss of face, there have been initiatives that did not live up to their promise or led up blind alleys.

Innovation

Let us return briefly to the mid-1980s, the start of what we described as a period of innovation. Again it was a time of change, perhaps like the mid-1960s. The purchase of John Wright, which with hindsight was perhaps an aberration, nevertheless illustrated a new spirit of adventure, of being willing to take risks. That was new to the Institute, at least on this scale, but from then on publishing turned out to be more adventurous and less averse to taking risks. The Vice-Presidents, Frank

Read and Eric Jakeman, supported the new approach. The change of senior management from Cecil Pedersen to Tony Pearce brought new ambition and a more competitive spirit.

The time was clearly ripe for new approaches, with fresh thinking in publishing, support from the honorary officers of the Institute, and the technological developments which were beginning to feature in publishers' thinking. Tony Pearce and his team took full advantage of these opportunities and set the scene for the next 10 years of growth, development and diversification.

One of the first indicators was the appointment in 1985 of Alan Singleton as Research and Development Manager. His sole brief was to explore opportunities for new business in all areas of interest to the Institute, from journals to green-field developments. Launching new journals into niches where there was a demand became a deliberate policy, and 24 new journals were launched or acquired between 1985 and 1997. Some have turned into successful and solid core journals, for example *Semiconductor Science and Technology* or *Waves in Random Media*; some were experiments which—at least in paper form—did not seem to meet a market need, such as *Engineering Optics* and *Liquids*, two reprint journals. Still others survive, successfully if modestly, to supply the needs of their particular communities.

Individual journal launches and their subsequent fates represent interesting publishing case studies in themselves. More pertinent for our purposes is the fact that, cumulatively, the journal launches during these years 1985–1997 substantially increased the Institute's market share of physics publishing and also took it into applied and interdisciplinary areas which were previously not accessible to it. *Smart Materials and Structures*, for example, or *Network: Computation in Neural Systems* were not natural outcrops of the kind of publishing represented by the *Journals of Physics* but took the Institute firmly into terrain occupied by other societies. Indeed, a number of new titles, including *Distributed Systems Engineering* with the Institution of Electrical Engineers, turned into partnership publications.

As we noted in connection with magazine development, the Institute's nervousness extended to new journals also. The reasons were similar: money was to be invested in publishing areas with which the Institute was less familiar, and despite the best business cases being presented to the publishing board, the risk was less easily assessed. Increasingly, however, the board became more comfortable with the notion that taking risks was part of publishing, and provided that, overall, the sums added up, it was acceptable to explore options which carried the risk of failure.

Failure is perhaps not the best way to describe our next interlude, for the story is more complex than that. The American Institute of Physics had long recognised the strength of Russian physics and capitalised on it through its journal translation series, with '*JETP*' being a flagship publication. It took the Institute (and other societies such as the Royal Society of Chemistry) a little longer to see the opportunity. When it did, it pursued opportunities in the USSR vigorously. Commissioning of books from authors in the region increased strongly. An editorial office was opened in the Ioffe Institute in what is now St Petersburg to channel papers to Institute journals and stimulate book proposals. Two new journals were launched: *Soviet Lightwave Communications* and the *Journal of the Moscow Physical Society*. Those were the days of *perestroika*, yet all was not to turn out well. In 1991 the Institute's USSR journal subscriptions collapsed. The two journals did not catch the Western imagination in author and subscriber terms and the Institute ceased to publish them in 1994. And as far as books were concerned, the recommendations of the Books Working Party led to a severe weeding out of titles originating from the region. The final blow as far as publishing was concerned was the collapse of the USSR and the subsequent decimation of scientific talent there.

Before we leave the innovation theme we need to acknowledge a number of developments that were initiated then and continue to prosper in different ways. The launch of *Physics World* was part of it, as was the launch of the *Physics World Buyer's Guide* (now *PhysicsNet*). So was a more flexible approach to the publication of meetings and conference proceedings, including the *IOP Short Meetings* series designed as a service to Institute groups and divisions. So was the attempt to launch a series of 'commercial' conferences and the start of the Institute's corporate affiliate programme. Both of these are now under the Institute rather than the publishing umbrella, with the corporate affiliate programme in particular associating some of the premier employers of physicists with the Institute's objectives.

Electronic publishing

We have only hinted at electronic opportunities so far in this account, principally because for 120 of the Institute's 125 years print-on-paper was the medium of choice. In the early 1980s talk was rife about the demise of print and the ascendancy of electronic processing and publishing. This turned to naught because computing power and bandwidth were not up to the requirements. It took another decade for technology to catch up with the perceived need for different approaches to some long-standing problems—information overload, for example, or inadequate funding in libraries.

At the Institute, electronic publishing started in the 1980s, when the use of wordprocessors for generating scientific papers became prominent. An internal survey of authors in 1987 demonstrated that TEX was the software of choice amongst physics authors at the time. The decision taken by the Institute—to make TEX and all its subsequent variants the standard in-house software route—had profound implications for future developments. It led to a build-up of deep expertise in text handling, first in a small expert unit led by Tony Cox and later spread throughout the Institute's production department. It provided consistency of approach while arguments raged about appropriate standards, including the now-favoured SGML. It left the Institute superbly equipped technically to take the step from electronic processing for print to electronic processing for electronic delivery.

Little wonder, then, that the Institute was the first physics publisher to publish a journal on the Internet: *Classical and Quantum Gravity Online* in 1994. Fairly basic it was, as a TEX file, but it was there! It took very little time to go the rest of the way, to a World Wide Web site in 1995 and the launch of the Institute's full electronic journals programme in January 1996, again ahead of other publishers. Unlike the print medium, in the electronic era nothing stands still, and the first version of electronic journals was soon embellished by extras such as multimedia enhancements, personalised filing cabinets and links from references to the INSPEC database of article abstracts.

Once 'up on the net', many other services followed. The Institute itself, including its branches, groups and divisions, featured strongly, as did services related to magazines and books. Developments have been very

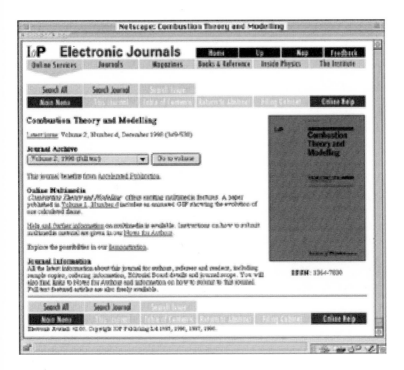

Publishing on the World Wide Web: the electronic version of 'Combustion Theory and Modelling' launched in 1997.

fast and there is enormous potential for exploiting the capabilities of 1990s personal computing and telecommunications to the benefit of individual physicists' information needs.

A question of growth

It remains to put some scale to the developments that we have chronicled over the last 125 years. In 1874 there was one journal, and the first volume contained 230 pages. A century later, in 1974, there were 11 journals containing 21,242 pages. By 1998 there were 34 research journals and 6 magazines, and about 50 books, with a total journal page extent of some 70,000 pages or thereabouts. Our sources do not reveal how many people were engaged in the publication of the

Exterior view of Dirac House.

Proceedings of the Physical Society, but we do know that 125 years later some 190 people are employed by the Institute's publishing company, mostly in the UK but also abroad.

In 1974 the Institute's journals were well supported by British authors, with only 54% of them from outside the UK. By the mid-1990s the overseas proportion was 85%. By that time a similar change was seen in the proportion of referees and of editorial board members. The circulation of the journals to libraries had changed from 15% UK to 9% over the same period. In 1874 the total revenue of the Physical Society was £506 0s 0d. It spent £36 0s 7d on 'printing' and £15 19s 2d on other expenses, leaving a surplus of £454 0s 3d. In the 1960s publishing was still a single line in the Institute's accounts. In the 1970s the contribution of publications to the net revenues available to the Institute increased and became of such size that the honorary officers of the Institute would take note. By the 1990s it began to dominate the Institute's finances. In 1974 total publishing turnover was less than £1m; by 1998 it was approaching £20m.

Monica Dirac at the official opening of Dirac House, named in honour of her father Paul Dirac, HonFInstP, on 23 September 1997.

Along the way the Institute survived the three-day week of 1974, with candles lighting the Victorian gloom of Lowther Gardens. It struggled through the hyper-inflation of the late 1970s (25% plus) and the consequent ravages done to the Institute's core activities, including redundancies at 47 Belgrave Square. Publishing moved offices repeatedly during the period. First it opened an overflow office in Bristol in 1971. Then it abandoned Lowther Gardens because its landlords, the Royal Commission for the Exhibition of 1851, increased the rent from peppercorn to market rate (a factor of 15) and it relocated all publishing to Bristol in 1975. Techno House, publishing centre from 1975 to 1997, was abandoned without too much regret in the latter year and Dirac House became the Institute's new publishing home.

Finally came two Queen's Awards for Export Achievement, in 1990 and 1995, acknowledging that in business terms the Institute's publishing had come of age and was able to measure up to the best.

CHAPTER 9

PHYSICS EDUCATION

On 18 November 1862, Faraday gave evidence before the Public Schools Commission, presided over by the Earl of Clarendon. He complained that the teaching of science had been almost completely overlooked in schools either as a matter of habit or prejudice. He told the Commission that 'science is now knocking at the door'.

> 'Men of science, fit to teach, hardly exist. There is no demand for such men. The sciences make up life; they are important to life. The highly educated man fails to understand the simplest things of science, and has no peculiar aptitude for grasping them. I find the grown-up mind coming back to me with the same questions over and over again.'

Although he deplored greatly the lack of science teaching, he made it clear to the Commission that he was not opposed to the teaching of classics, but he regretted that those who had had a classical education, persons who had been fully educated by the existing system, were ignorant of their ignorance. He enlarged on what he believed should be the object of education, namely:

> 'to train the mind to ascertain the sequence of a particular conclusion from certain premises, to detect a fallacy, to correct undue generalisation, to prevent the growth of mistakes in reasoning. Everything in these must depend on the spirit and the manner in which the instruction itself is conveyed and honoured. If you teach scientific knowledge without honouring scientific knowledge as it is applied, you do more harm than good. I do think that the study of natural science is so glorious a school for the mind, that with the laws impressed on all these things by the Creator, and the

wonderful unity and stability of matter, and the forces of matter, there cannot be a better school for the education of the mind'.

Without doubt science played a very small part in school education and then almost exclusively in the public schools and certain endowed grammar schools. Some schools had high reputations, for example Clifton College had a Fellow of the Royal Society, W H Shenstone, as senior science master, for over thirty years; two other members of the staff were to become Fellows of the Royal Society and a laboratory assistant there became, in due course, Sir Richard Gregory, another FRS and the editor of *Nature*.

Help for teachers, particularly in London, came from the Physical Society after its formation in 1874 when regular fortnightly meetings were held at 3 o'clock on Saturday afternoons from November to June in the Royal College of Science. There was always great interest in demonstrations and experiments, especially in any new developments.

Sir William Huggins, at the anniversary meeting of the Royal Society in 1902, devoted his presidential address to the failings of schools. He saw them as the cause of the lack of appreciation 'of the supreme importance of scientific knowledge and scientific methods, especially on the part of the leaders of the nation'.

The Association of Public School Science Masters, formed in 1902, became in December 1918 the Science Masters' Association (SMA). Eventually the SMA was to combine with the Association of Women Science Teachers (AWST) to become the Association for Science Education (ASE). These associations had a powerful influence on physics education throughout the 20th century and both the Physical Society and the Institute of Physics maintained a close liaison with these associations.

However, the content of the physics syllabuses in use in secondary schools reflected, throughout the next 50 years, physics prior to 1895 The subject matter was exclusively classical and divided into the distinct compartments of mechanics, heat, light, sound, electricity and magnetism. Most of the teaching was directed towards either the School Certificate or the Higher Certificate examinations, and relied much on rote memory of formulae, definitions and the description of routine experiments.

The 1950s and 1960s

After the end of the Second World War there was concern that physics syllabuses needed to be updated and the Institute was represented on a committee, set up by the science teachers' associations, to produce a report, *Physics for Grammar Schools*, on what should be done. A member of the committee was a young physics teacher, John Lewis.

It was soon realised that there needed to be some material representing physics later than 1895 included in the syllabus and a new Committee on the Teaching of Modern Physical Science was set up to develop apparatus and to suggest methods of teaching. The Institute was involved through Professor H Lipson, L R B Elton and Norman Clarke, and John Lewis was invited to be Chairman. Its strength lay in a number of innovative teachers who were supported and encouraged by the Institute and the Association for Science Education. It was also significant that HM Inspectorate accepted membership of the committee and took part in its activities: there was a great sense of working together and much new experimental work was developed.

Curriculum development work was started in the United States following the launch of the first Sputnik and the programme produced by the Physical Science Study Committee (PSSC) had an influence throughout the world. Despite the good material within the course, it was written as a one-year course for American schools to suit the American education system, and in consequence never caught on in European countries which had traditions of physics courses extending over many more years. There was another influence: an extended tour by John Lewis in the USSR involved a study of physics teaching in Russian and Ukrainian schools. He was not impressed by the experimental work being done, by the apparatus or the films available for class use, but was by the organisation behind the teaching. They claimed both in Russia and the Ukraine that they had enough physics teachers, but that they would never have enough good teachers, and that therefore they must provide the where-with-all to enable an indifferent teacher to achieve a standard. No school was opened without the necessary apparatus, films or other available equipment, but above all the teachers were provided with detailed teachers' notes and given guidance on how to teach. By contrast, teachers in Britain at that time would receive a degree in physics and then would teach as they were themselves taught for the next 30 years with a syllabus as the only guide.

The Director of the Nuffield Foundation learned of Lewis' report. He arranged a dinner at Nuffield Lodge with Lord Hailsham, Minister for Science, David Eccles, Minister of Education, Sir Alexander Todd, Sir John Cockcroft and others at which John Lewis spoke. The outcome was the setting up of the first of the Nuffield Projects. Until that time the work of the Foundation was mainly devoted to medical work and this marked the beginning of its major commitment to education. A National Committee on Physics Education was set up jointly by the Royal Society and the Institute under the chairmanship of Sir Nevill Mott to consider all aspects of physics education in the UK. It decided first to direct its attention to physics teaching in secondary schools and thereby became involved with the Nuffield work, and Sir Nevill Mott became the Chairman of the Nuffield Physics Projects. The 11–16 Physics Project, led initially by Donald MacGill from Scotland and, on his untimely death, by Professor Eric Rogers with John Lewis and Ted Wenham, was followed by the A-Level Project led by Paul Black and Jon Ogborn.

The impact of the Nuffield Physics Projects was profound as they influenced all physics courses in the UK. Syllabuses previously compartmentalised into heat, light, magnetism and electricity disappeared and unifying themes like waves and fields began to appear. Many of the routine experiments also disappeared, there was a greater interest in investigations, and students learnt to look for evidence. There was much less rote memory, pupils were encouraged to think for themselves, and there was a lot more fun in physics teaching.

The Institute began to have heavy involvement in physics education internationally. Norman Clarke was secretary of the first international conference on physics education, held in Paris in July and August 1960. This stimulated interest in education internationally and one consequence was that the International Union of Pure and Applied Physics set up an International Commission on Physics Education with Norman Clarke as Secretary and with which the Institute remained closely associated for the next 30 years.

The 1970s and 1980s

In the two previous decades the Institute had contributed to major curriculum changes culminating in the Nuffield A-level project. The next two decades marked the considerable expansion of the Institute's Education Department, with help and encouragement being given directly to schools and to teachers.

The Institute initiated and serviced from 1973 onwards the Joint Royal Society/IOP Committee for Physics Education, the Standing Conference of Professors of Physics, the IOP Education and Careers Committee, the Joint Committee for HNC/HND in Applied Physics, the Education Committee of the European Physical Society, and from 1977 the Committee of Heads of Physics Departments in Polytechnics.

A wide range of working parties produced reports and booklets on:

- *Girls and Physics*
- *The Shortage of Physics Teachers*
- *Interface between Physics and Mathematics*
- *Interface between Physics and Engineering*
- *A-Level Core Syllabus in Physics*
- *Resources for Teaching School Physics*
- *Strategies for Physics in the UK*
- *Form and Content of Sixth Form Physics*

and others.

The Education Department was responsible for the Institute's Graduateship Board of Examiners. This was highly regarded as a qualification and it gave a valuable opportunity for mature students to obtain a degree in physics. During the time of its existence from 1952 to 1984 there were 3261 candidates of whom 1203 reached honours degree level.

The Institute's acceptance of the importance of vocational and technical education was shown by its chairmanship of the validating panel for higher level BTEC courses in physics.

The Institute had for some time been monitoring, and sounding an alarm on, the deteriorating situation for physics in secondary schools, especially the shortage of graduate physics teachers. The Department of Education and Science had belatedly recognised the seriousness of the problem—and a government spokesman publicly acknowledged the part played by the Institute of Physics in achieving this recognition.

The increasing level of resources made available to the Education Department allowed support to teachers and students to be extended in several ways:

- During this period, up to eight **Physics at Work exhibitions**, each lasting several days, were held annually throughout the UK. At these exhibitions students were shown the many different ways in which physics was used in industry.

- A series of **annual school lectures** with titles such as *Musical Squares*, *Science of Soap Films and Bubbles* (Dr Cyril Isenberg), *Say it with Frozen Flowers* (Dr Mervyn Black), and *Serpents and Synthesisers* (Dr Murray Campbell). Each was given in up to 30 venues throughout the UK with total audiences of up to 10,000.

- The encouragement by the Institute of exceptional teaching of physics was shown by the **Teachers of Physics awards** which were given annually to five secondary school teachers who were recognised as having inspired their pupils by the excellence of their teaching, and on many occasions an award was extended to a primary teacher known to have done the same in teaching physics or science.

- The first three-day **Physics Update Course** was held in Malvern College for secondary school physics teachers to allow them to become acquainted with the latest ideas in physics and physics education. Such courses, run by the Ministry of Education, had been regular features for physics teachers in the 1950s and 1960s, but they had all lapsed. Such was the success of this first course that from then onwards an update course was always held annually in Malvern College and in two different universities around the country. Support for these courses was generously given by the Armourers and Braziers Company.

- For several years the **Small Grants Scheme** allowed grants of up to £1000 to be given to physics teachers to allow development

*The 1998 Annual Schools' Lecture on 'Black Holes, Wormholes and Time Travel'
given by Jim Al-Khalili of Surrey University.*

projects to be run in schools and, as a result, a substantial number
of interesting projects resulted.

- Another innovation was the **Schools' Affiliation Scheme**, in
  which, in return for a small subscription, schools were able to
  be more closely linked with the Institute and with practising
  physicists. Over 600 schools became involved in the first year
  of its launch and the figure soon exceeded 1000. The scheme was
  particularly useful for younger and less experienced teachers or
  teachers who did not hold qualifications acceptable for professional
  membership of the Institute.

- The increased resources also allowed the range of publications to be substantially increased. An informal newsletter, *Snippets*, with short pieces of interesting physics and a variety of ideas was produced. It was issued termly and sent to every secondary school in the UK. Such was the interest it generated that in due course it was also sent to all members of the Institute. Booklets on *Physics Courses in Higher and Further Education* were produced. There were a number of booklets providing statistics on aspects of physics and physics education, and there was a biennial salary survey conducted by the department, all of which provided valuable information for the physics community at various levels.

- A video called *Physics All Around Us* was issued during 1987 and won a first prize in the education section of the British Association of Film and Television Awards.

Of particular importance in 1988 was the document *Physics In Higher Education* which was prepared for Council by a working group and widely circulated to vice-chancellors of universities, directors of polytechnics, government departments, members of Parliament, the University Grants Committee, and others. It received wide attention with numerous comments in the media and was included as an annexe to the review *The Future of University Physics* prepared for the University Grants Committee by a working group chaired by Professor S F Edwards.

These two reports led in due course to the introduction of the four-year MPhys/MSci degrees in all major physics departments. Such degrees were intended for professional physicists and complemented the three-year BSc, which was also changed to become a degree for individuals who would most probably follow careers outside physics.

Advancing Physics: the Institute's 16–19 Physics Project

The 1960s saw radical changes in physics education. The Nuffield Projects had attempted to bring the content of courses up to date and to suggest a new style of teaching and learning. There were further developments in the following decades, for example the Science and Technology in Society (SATIS) work, which aimed to show the relevance of science to social, economic and environmental issues.

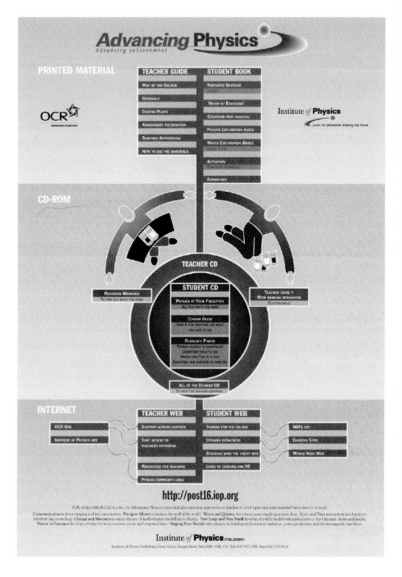

'Advancing Physics', the 16–19 Physics Initiative.

By the 1990s there was an inadequate number of students taking A-level physics examinations. Our national economy depends on a flow of high-calibre scientists and engineers, and physics is an essential part of their education. There was an urgent need to attract more young people to study physics through high-quality courses which excite and challenge them. The Institute realised that more was needed than the revision of syllabuses which examining boards could offer. For this reason, the Institute of Physics decided to launch a new 16–19 Physics Initiative to make physics courses more exciting, up-to-date and relevant. A team was set up in 1997 under Professor Jon Ogborn so that something radical could be achieved. Just as the Nuffield Projects made their powerful influence in the 1960s, no doubt there will be a similar or greater impact on physics education in the early years of the new millennium as a result of this initiative.

CHAPTER 10

PHYSICS AND PROFESSIONALISM

A profession is defined in the *Oxford English Dictionary* as 'a vocation in which a professed knowledge of some department of learning or science is used in its application to the affairs of others or in the practice of an art founded upon it'. The term seems to have been used in Britain as early as the 12th century for those in a religious order. By the 16th century the three learned professions of divinity, law and medicine were recognised, as also was the military profession.

It was realised at an early stage that some sort of regulatory institutions were needed in order to maintain the standards of the professions and it became common for these to have the authority of Royal Charters. The Roman practice of using written charters to confirm and record grants of land was introduced into Britain by Theodore, Archbishop of Canterbury from 669–690 AD. The 'Charter of Liberties', by which the king on his accession undertook to rule by the law, was introduced by Henry I in 1100 and led up to *Magna Carta* in 1215. Royal Charters were then used to define the rights and privileges of institutions such as towns, abbeys and colleges. Later, with the growth of Empire, companies were added to the list. The British East India Company received its Charter in 1600 and the Hudson Bay Company in 1670.

The first grant of a Royal Charter to a professional institution seems to have been by Henry VIII to the Royal College of Physicians in 1518. Under its Charter, the College lays down the rules to be obeyed by those who practise medicine and sets out the duties and privileges of the profession. The Law Society received its Charter in 1831. The Institute of Chartered Accountants may have been the first to use the word 'Chartered' in its title: it received its Charter in 1880, though in Scotland accountants called themselves chartered and put the initials 'CA' after their names some 15 years earlier. The chartered institutions are now answerable to the Queen's Privy Council and are empowered to award chartered professional qualifications.

Engineering institutions

In about 1771 John Smeaton, Robert Stevenson, James Watt and others established 'The Society of Civil Engineers'—otherwise known as 'The Smeatonians'. Civil engineering was defined, of course, to be all engineering that was not military in nature, and therefore encompassed the engineering disciplines then practised. The purpose of the Society was expressed as follows:

> 'That it would be well if some sort of occasional meeting, in a friendly way, could be held, where they might shake hands together and be personally known to one another; that thus the sharp edges of their minds might be rubbed off, as it were, by a closer communication of ideas, no ways naturally hostile'.

The Smeatonians met at the Palace of Westminster, the Courts of Justice or the King's Head in Holborn. The Society was briefly wound up in 1792, but reformed in 1793, meeting at the Crown and Anchor in the Strand where

> 'members might dine together at a late hour, after attendance at Parliament, and pass the evening in that species of conversation which provokes the communication more rapidly than it can be obtained from private study or books alone'.

The rules were progressively developed and the Institution of Civil Engineers was established in 1818 and incorporated by Royal Charter in 1824. Membership was defined to include:

> 'Ordinary members—real engineers actually employed as such in public or private service. Honorary members—men of science and gentlemen of rank and fortune, who had applied their minds to subjects of Civil Engineering, and who might, for talents and knowledge, have been real engineers, if it had

not been their good fortune to have it in their power to employ others in this profession'.

By 1847 its membership had reached 615, but it was not a welcoming place for those who saw themselves as natural scientists.

The study of electricity was becoming popular, although not as a topic for the Institution of Civil Engineers. In 1830 there was established an independent organisation involved in science known as The Adelaide Gallery of Practical Science. Charles Wheatstone carried out experiments there in 1834 to determine the velocity of electricity with a rotating mirror. In 1838 Faraday determined the character of the shock of the gymnotus (sic).

'Clever professors were there, teaching elaborate sciences in lectures of twenty minutes each.'

In 1837 the Electrical Society of London was established by the efforts of William Sturgeon and it met in the Adelaide Gallery, which had now become 'The Royal Gallery of Practical Science'. Many famous names were associated with the Society in this period, including Joule. Sadly it ran up a debt of £85 as a result of its publishing activities and it was wound up in 1843. Sturgeon kept the flag flying, and he edited and produced *The Annals of Electricity, Magnetism and Chemistry* and *Guardian of Experimental Science*. The pressure to re-establish the society grew over the coming years, and finally in 1871 the Society of Telegraph Engineers was formed as a result of the rapid development of telegraph cables. By that time there was a transatlantic cable, as well as cable links between Malta and Alexandria, France and Nova Scotia, and Suez and Bombay. The new Society attracted engineers engaged in these endeavours, as well as scientists active in all the related fields of electricity.

Meanwhile, the Institution of Civil Engineers was losing members, mainly those who were engaged in other fields of engineering. In particular, the Institution of Mechanical Engineers was formed in 1837. Some say that this was the result of the Institution of Civil Engineers being reluctant to confer membership upon George Stevenson. Whatever the reason, it became the pattern that newly emerging branches of science and technology spawned a new society.

The Society of Telegraph Engineers went on to become the Institution of Electrical Engineers. Thereafter societies and institutions were regularly formed, and occasionally wound up. By 1979 when the Finniston Report was published there were close to 50 engineering bodies.

In some ways the engineering institutions have set an example in providing professional qualifications which has influenced the Institute of Physics. The first engineering body to be chartered was the Institution of Civil Engineers and it considered itself as representing all engineers and it attempted to prevent others, including the mechanicals and the electricals, from forming separate institutions. The Institution of Electrical Engineers applied for a Royal Charter in 1880, but it was blocked by the civils. It was finally granted its Charter in 1921. In 1924 the Privy Council granted the right for corporate members of the IEE to describe themselves as 'Chartered Electrical Engineers' and this became the accepted usage. The Institution of Mechanical Engineers hesitated in the early 1920s to become chartered as any changes to the constitution or bylaws had to go through the Privy Council, but by 1930 it changed its mind and received its Royal Charter, after which members could call themselves Chartered Mechanical Engineers.

In later years, as engineering diversified and other institutions were set up, it became clear that some moves to unify the engineering profession were necessary. After many years of discussion between the various institutions, the Council of Engineering Institutions (CEI) was set up by Royal Charter in 1965 and the title Chartered Engineer (CEng) was established. The Finniston Report, published in 1980, stated:

'The professional status of engineers is not generally acknowledged in this country by the public and by employers to the extent that it is for, say, doctors or lawyers'.

This led to the replacement of the CEI by the Engineering Council with a Royal Charter granted in 1981. The standing of CEng as a professional qualification has now become generally recognised and is often stated as a requirement in job advertisements.

One result of these developments has been that the engineering institutions in recent years have become very concerned with the professional training of engineers and this has led to the accreditation of degree courses and continuing professional development and other aspects of training.

The Institute of Physics

When the Physical Society was formed in 1874, its main concern was with the reading and discussion of communications in physics. The members were mainly academics and their interests were in research. It was primarily a learned society. But by the end of the First World War, physicists were working professionally in other fields, particularly in industry and government service, and they felt the need for an organisation that would look after their interests as professionals, and thus the Institute of Physics came into being as an independent body. In due course—as described elsewhere in this book—the Society and the Institute were merged in 1960 and from then onwards operated both as a learned society and as a professional body with interests in (i) education, (ii) professional qualifications and (iii) services to members. Other professional concerns involved: representations to government or other bodies on issues of concern to physicists; the promotion of physics in the community; relationships with overseas bodies.

Professional qualifications for physicists

From its founding in 1919, the Institute of Physics attached considerable importance to the professional status of its members in their grades of Fellow, Associate and Ordinary member. It detailed the qualifications needed for entry into each class of member, issued diplomas and maintained a register of members.

After the amalgamation of the Institute and Society, a distinction was made between those with the professional grades FInstP, AInstP and GradInstP, and those which were not professional, which included 'Fellows of the Former Physical Society', although these were invited to apply for professional status if they wished.

The most significant step towards professionalism was made when in 1968 it was decided to apply for a Royal Charter—see pages 124–126 in chapter 7—and this was granted by the Queen on 6 November 1970,

The Institute was awarded 'nominated body' status by the Engineering Council in 1996: Brian Manley (President, 1996–1998) is pictured here with Mike Heath, Director-General of the Engineering Council.

thereby confirming the Institute in its new chartered status as the pre-eminent body representing physics and physicists in the UK.

Following the general recognition of Chartered Engineer (CEng) as awarded by the Engineering Council, the question was raised whether the Institute should embark on something similar. From 1982 to 1984 the Institute's Council had lengthy discussions on this subject. In the end it was decided to approach the Privy Council to get approval for the use of the title 'Chartered Physicist'. In March 1985 it was announced that the Privy Council had approved the change in the Bylaws which would allow the title 'Chartered Physicist' to be used. Since then, corporate members have been able to use this title and to put 'CPhys' after their name.

It was also decided to discuss with the Engineering Council the possibility of members of the Institute with appropriate experience being allowed to become Chartered Engineers. In 1985 a panel from the Engineering Council visited the Institute and recommended that the

Institute should become an 'Institution Affiliate' of the IEE, which was a nominated body to approve applicants for CEng. This took place and several hundred members of the Institute have been given CEng status by this route. Following reorganisation in the Engineering Council in the mid-1990s, the Institute was given 'nominated body' status to make its own recommendations direct to the Engineering Council.

After the Second World War the Institute felt that the educational facilities available did not meet all the needs of professional physicists and in consequence it set up its own examinations. In 1952 the Institute's graduate examination was started to provide an academic qualification equivalent to a degree for those who had not had the opportunity to gain the honours degree needed for professional status. Courses for it were provided, with the Institute's approval, by a number of polytechnics and colleges of technology. It continued until 1984 when dwindling numbers of applicants led to the courses being dropped.

At a less advanced level, there was another route to professional physicist through part-time courses at a technical college. It was intended for those who left school at 16 and went into employment. After taking a Higher National Certificate (HNC) examination, they would enter a course which would qualify them to take the Institute's graduate examination and thus achieve full status. HNC and Higher National Diploma (HND) examinations continued to be held until 1983, the year before the termination of the graduate examination.

Other professional activities

In the 1990s the Institute provided an increasing variety of services to members including advice on qualifications and career development, information on employment opportunities and related matters. Professional briefs were issued in the form of booklets giving details on 'job seeking', 'consultancy work', 'training for women' and other matters. The Benevolent Fund continued to be maintained to assist members or their dependants who fell on hard times.

From 1991 the Council of the Institute agreed to sponsor a Professional Development Scheme which would assess employers' training schemes and encourage young physics graduates to acquire the necessary training and experience to qualify them for Chartered Physicist status. In

1996 the Institute's Continuing Professional Development scheme was started, recognising that throughout a career professional physicists needed to continue learning and training activities.

In the late 1990s it had become increasingly clear that, if the Institute was to preserve its professional status, it had to establish and maintain links with business and industry in physics-related areas.   It had representatives in a number of industrial locations to ensure that its links with industry were maintained. In 1989 the Corporate Affiliate Programme was started and as the millennium approached there were 25 organisations within the scheme. They were represented on a number of Institute boards and committees and engaged in a variety of other related activities.   In 1995 the Institute launched the SME Club to cater for the special interests of small and medium-sized enterprises. The Institute published a number of booklets containing information of particular interest to industry and in 1996 it started regular publication of the newsletter *Physics in Business*. All these and similar activities gave the Institute a well deserved reputation as a supporter of industry.

It was increasingly recognised that there is a need for the public to be made aware of the vital role physics plays in contemporary life. In consequence a Public Affairs Department was set up in January 1991 and this established links with the media to promote physics.   The Institute has always had links with physicists in other countries and continued to play a part in setting up overseas organisations and giving encouragement to others. For example, the Institute played a major part in setting up in 1994 and maintaining thereafter a European Register of Physicists.

CHAPTER 11

THE 1990S AND BEYOND

The arrival in 1990 of a new Chief Executive, Dr Alun Jones, and a new President, Professor Roger Blin-Stoyle, led almost immediately to the provision of a corporate strategy for the five years 1991–96. To satisfy the Institute's Charter, three targets were agreed:

- To achieve a net increase of 5000 members over the five-year period;
- To provide new and improved services for Institute members;
- To launch a *Campaign for Physics* to ensure that physics not only survived, but flourished.

Reorganisation of the Institute's internal management was agreed by Council and took effect from 1 January 1991. Four new directorates were set up, each with its own director responsible to the Chief Executive and with objectives in line with these targets. These directorates were: Education, Industry and Public Affairs; Publishing (which already existed); Qualifications and Membership; and Services to Members. A Director of Finance was appointed to be responsible for both the finances of the Institute as a whole and the finances of the publishing company (a wise move, as Institute finances by then depended heavily on publishing).

Membership matters

At the end of 1990, Institute membership was 14,723, so aiming to reach 20,000 by 1996 was considered ambitious. Happily, not only did the number of members reach 20,212 by 31 December 1993, but the number of Chartered Physicists also reached an all-time high of 10,915.

Alun Jones, Chief Executive of the Institute.

The Institute's recruitment campaign was greatly helped by the Privy
Council's agreement in autumn 1993 that the Associate Member class
could be divided into 'Associate' and 'Graduate Member', with the
designatory letters GradInstP being granted to the latter.   To be a
Graduate Member, individuals needed an academic qualification which
met the requirement of corporate membership, so the new designation
provided an added incentive for student members to continue with the
Institute on graduation. Membership continued to rise to 21,674 by the
end of 1995, with a corporate membership of 11,370.

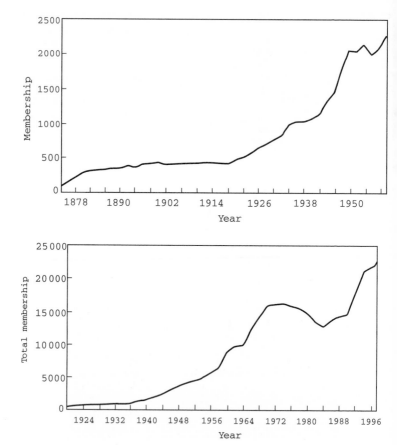

Membership growth: the Physical Society 1874–1960 (top) and the Institute of Physics 1920–1998 (bottom).

In 1993 the female membership of the Institute increased by 67% and numbered over 2000. Professor Daphne Jackson bequeathed funds to the Institute to encourage women physicists to return to employment after a career break and a Trust was established for this purpose.

In 1992 a student working party was set up and this evolved into NEXUS, a network of student physical societies. In 1999 NEXUS continued to issue a regular newsletter and organised meetings of society representatives. It also helped the Institute (which now had a Student Liaison Officer on its staff) to develop a strategy for attracting and retaining student members.

International affairs

The chartered Institute has always had an international role and in the 1990s contact with other institutes of physics and physical societies worldwide led to a number of joint cooperative agreements. In 1999, there were agreements with:

American Physical Society
Australian Institute of Physics
Canadian Association of Physicists
French Physical Society
German Physical Society
Indian Physics Association
Japan Society of Applied Physics
Physical Society of Japan
New Zealand Physical Society
Polish Physical Society
South African Institute of Physics.

In three instances, a tangible reminder of these comes in the form of prizes or awards (see also appendix G). The first such prize was instituted in 1945 jointly by the French and British Physical Societies as a memorial to Fernand Holweck, Director of the Curie Laboratory of the Radium Institute in Paris, who was killed by the Gestapo during the occupation of France. In 1972 the Institute of Physics and the German Physical Society established an award to commemorate Max Born and to cement the link between the two societies. The third joint award is the Harrie Massey Medal and Prize, established by the Institute of Physics in 1988 to commemorate the 25th anniversary of the founding of the Australian Institute of Physics.

An annual exchange of lecturers between the Institute of Physics and the Indian Physics Association was instituted in 1997. The visiting Indian lecturer to Britain and Ireland is called the Homi Bhabha Visiting

Lecturer and, in the alternate year, the lecturer to India from Britain or Ireland is known as the Cockcroft and Walton Lecturer.

By the 1990s the European Physical Society was firmly established. Founded in 1969, its aims continued to be:

- To provide a forum for the discussion of subjects of common interest to all European physicists;
- To provide a means for action on matters which cannot conveniently be handled by national bodies (such as the setting up of an information bulletin or a European journal);
- To coordinate the organisation of scientific meetings, summer schools and the like.

The Institute of Physics joined EPS as a founding member society and continued to play a leading role in its development in the 1990s. Members of the Institute are automatically national society members of EPS, which represents over 70,000 members in Europe. As the millennium approached, EPS publications included *Europhysics News*, *EPS Conference Abstracts* and two journals, *European Journal of Physics* and *Europhysics Letters*. Six divisions each covered their own sectional interests and interdivisional groups covered Accelerators, Applied Physics and Physics in Industry, Computational Physics, Experimental Physics Control Systems, History of Physics, Physics Education, and Physics for Development. In the mid-1990s the Institute was associated with the EPS's European Mobility Scheme for physics students and made a major contribution to the establishment of a European Register of Physicists which awards the professional qualification 'European Physicist' (EurPhys).

On the campaign trail

One thrust of the 1991 corporate strategy plan was to ensure that physics not only survived but flourished. *The Campaign for Physics* was launched with a press conference at headquarters in May 1992, with a separate Scottish launch in June. These occasions were used to publicise the Institute's new report on physics-based industries (PBIs), which provided quantitative support for anecdotal evidence that PBIs were thriving and making a major contribution to the national economy. Considerable media attention resulted.

In 1994 the Office of Science and Technology (OST) and the Department of Trade and Industry launched OST's 'Technology Foresight' programme, which aimed 'to identify areas of research and technology likely to meet perceived market requirements over the next 10–20 years'. The Institute was anxious to ensure that the importance of physics was fully recognised and nominated members to sit on most of the subject panels proposed by OST; several were appointed. To ensure that physics was given full consideration, the Institute set up nine 'shadow' panels from within its own membership in areas where the application of physics was considered to be specially relevant. The output of these shadow panels was coordinated centrally and the booklets produced were forwarded to OST and its panels as well as being made widely available. The Institute clearly showed its intention to do all it could to press the case for physics and physics-based industry through the Foresight programme.

On the arrival of a new research council structure in 1993, the Institute expressed its concerns about the role of fundamental physics and the importance of assessors representing the physics community on both the Particle Physics and Astronomy Research Council (PPARC) and the Engineering and Physical Sciences Research Council (EPSRC). The first steps to implement government policy as outlined in the White Paper *Realising Our Potential* led to fears that basic physics might not be given the support it deserved. The Institute obtained assurances from both the Cabinet Office and the chief executive of EPSRC that basic science was an important component of the new policy and that support would continue to be provided. It nevertheless monitored the situation in order to react to the experiences of academic physicists dependent on research council support.

As the millennium approached, the Institute took every opportunity to press the case for physics and physicists, as witnessed by a long list of evidence given, responses made and correspondence exchanged with government departments and other official bodies since the early 1990s. The Institute's profile as a body concerned with policies and legislation was raised to a high level by its presence at press conferences and parliamentary select committees, as well as by the provision of contributions—both invited and uninvited—to parliamentary enquiries. The Institute's many communications in the 1990s established a close rapport with the OST which will prove advantageous in the years ahead.

The 1990s also saw the first issue of *Physics in Business*, a thrice-yearly newsletter targeted at the Institute's members in industry. Increased recruitment within the industrial sector was, in part, due to the innovative and improved services which the Institute offered.

Professional developments

The new directorate for Services to Members uncovered a need for short courses on professional development for physicists working outside of academe. Other initiatives included new professional groups covering consultancy and total quality management, as well as the provision of discounts on financial, insurance and health services that proved popular with the membership at large. A new computer-based job-matching service was introduced in 1993; it made impressive progress and was used by 320 members in its first year. A 16-month diary was introduced and continues to be issued to all members each September; it includes information on the Institute as well as listing contact addresses and data on physical constants.

A series of professional and legal briefs was written and made available to members: *Obtaining Grants and Awards*; *Professional Training for Women*; *Intellectual Property Rights*; *Working Abroad—European Community*; and *Starting as a Consultant*.

Education for all

Developing, integrating and promoting the Institute's educational, industrial, public awareness and public relations activities were given a high priority in the 1991–96 strategy plan.

Educational activities aimed at schools and colleges included the publication of the *Snippets* newsletter (superseded by *Phases* in 1998), 'Physics at Work' exhibitions around the country, and a small grants scheme for teachers. The annual series of schools lectures continues to be given in up to 40 venues and excellence in science and physics teaching is recognised through annual awards. Colourful, lively posters and several videos have also been produced. 1996 saw the tenth 'Update' course for teachers held at Malvern College and throughout the late 1990s Update courses continued to be held in two different universities around the country.   There was also important work monitoring what was happening in schools, in particular the long-standing shortages of physics teachers (especially in the lower levels in secondary schools where, more often than not, physics has been taught by biologists).

The Institute's focus on public awareness of physics revealed the need to have physicists ready and willing to 'meet the media'. Media representatives were appointed in the branches and a *Media Representatives' Handbook* was produced.   This resulted in greatly increased press coverage throughout the 1990s, including a series of posters displayed on London Underground trains in 1998. A *Physics and Fine Art* video was produced with the National Gallery.

Conferences galore

In a typical year in the late 1990s the Institute's Conferences Department was responsible for between 30 and 40 one-day conferences, 10–15 large conferences lasting up to four days as well as several international conferences.   The annual Condensed Matter Conference attracted up to 1000 participants whereas one-day conferences typically welcomed between 30 and 100 physicists.

A significant innovation was the introduction of an Annual Congress, the first of which was held in Brighton in April 1993. Almost 1000 people attended and 100 commercial companies were involved in the associated exhibition. The plenary speakers were the Rt Hon William Waldegrave MP, Chancellor of the Duchy of Lancaster and the UK's Science Minister at the time, and Professor Allan Bromley, President Bush's Science Advisor.   Subsequent congresses in the 1990s were held in Brighton (twice), Telford (twice), Leeds and Salford. A broad spectrum of conference topics and trade exhibitions generally attracted 800–1200 participants.   As Congress developed, more components aimed at junior and secondary school pupils were introduced. At the end of the 1990s, these 'Physics in Action' events continued to be highly successful, attracting some 2000 children and usually entertaining the then current Science Minister who was usually delighted to accept an invitation to attend.

The 1998 Annual Congress at Brighton: Brian Manley (President, 1996–1998) is pictured here with John Battle MP, Minister for Science, Energy and Industry.

A new home

The most significant event of the 1990s was the move of the Institute's headquarters. After a long search, 1994 saw the Institute exchange contracts for the acquisition in mid-1995 of a long lease on new premises at 76 Portland Place, London W1. 47 Belgrave Square, which had been the Institute's home since 1947, had become increasingly unsuitable for both members and staff.

Exterior views of 76 Portland Place (the lower photograph is of the President, Gareth Roberts).

The new building houses a 170-seat lecture theatre and conference centre, together with greatly improved member facilities including a club room, a work room with IT services and a library, as well as a complex of committee and meeting rooms. In addition, the staff was accommodated in more modern premises which did not suffer from the labyrinthian disadvantages of 47 Belgrave Square. The Institute left Belgrave Square with more than a tinge of sadness, but progress in physics and the increasing number of members demanded modern facilities and up-to-date services for the 21st century. Moving to Portland Place was the largest project which the Institute had ever embarked upon and probably the most significant in terms of providing support for physics and physicists.

The project cost £5.7m, of which £3.1m was paid for a 60-year lease and the remainder was the cost of building works, principally the construction of the new Rutherford Conference Centre. This was met from the generous benefaction in the late 1980s of £1.7m from the will of Major C E S Phillips, Honorary Treasurer of the Institute from 1925–45 (a benefaction which came to the Institute after the death of Major Phillips' widow), from the sale of the lease of 47 Belgrave Square (£0.8m) and from surpluses generated by the Institute's publishing activities.

Her Majesty visits

The Institute was greatly honoured to have the Queen open its new headquarters on 11 December 1996. Her Majesty was greeted on arrival by the President, Dr Brian Manley, past Presidents Dr Clive Foxell and Sir Arnold Wolfendale, and the Chief Executive, Dr Alun Jones. She was presented with a posy by the Institute's youngest headquarters employee, Miss Katie Knight. As well as staff, the audience comprised members of Council, several other past Presidents and their wives, and representatives of the contracting companies and professionals concerned with the acquisition and refurbishment of Portland Place and the construction of the Rutherford Conference Centre.

After the unveiling ceremony Her Majesty was escorted through an exhibition showing the role and importance of physics in the community.   A stand describing the activities of the Institute incorporated an Internet link to demonstrate the Institute's electronic

HM The Queen with Brian Manley (President, 1996–1998) at the official opening of 76 Portland Place on 11 December 1996.

publishing activities.   Her Majesty was intrigued to see that a photograph of her taken by digital camera when she arrived was already on the Institute's home page on the World Wide Web by the time she arrived on the stand 25 minutes later.

Her Majesty and the Royal Party were entertained to tea, during which she met invited guests and members of staff. As she left—considerably later than planned—Her Majesty thanked those who had organised the visit.

A current snapshot

Summarising the Institute's position at the end of the 1990s is a difficult task but perhaps a valuable picture can be painted by listing what happened at the 1998 Annual Congress in Brighton:

- Four plenary lectures—two of them by Nobel Laureates and one by a government minister;

The Rutherford Conference Centre at 76 Portland Place.

- The Congress Debate, led by Sir Hermann Bondi and Sir Arnold Wolfendale, which stimulated a substantial and lively discussion on the future funding of physics;
- Evening public lectures which attracted over 1000 people;
- Lectures and hands-on activities for some 2500 schoolchildren;
- 14 scientific conferences arranged by the Institute's groups and divisions;
- *Physics World Expo*, a substantial commercial exhibition;
- A Guest Programme which brought politicians, industrialists, academics, scientists and others to Congress to experience the breadth and depth of physics;
- Technology seminars run in conjunction with the exhibition, which exceeded the expectations of the organisers;
- Short courses on technical subjects aimed at a range of scientists and engineers from industry and academia;
- A meeting for the Institute's Corporate Affiliate companies;
- Lecturing and poster competitions for undergraduate and postgraduate students arranged by the Institute's student section, NEXUS;
- A meeting of the Institute's Council.

Delegates, the exhibiting staff, course attendees, the general public and schoolchildren totalled over 5000 individuals. In addition to those actually present, coverage in the press and television resulted in many more people hearing about the Institute's Annual Congress and the role that physics plays. The week began with four- and five-column articles in *The Times* and the *Daily Telegraph* on subjects dealt with at Congress, and continued with coverage from BBC News, Canadian Broadcasting, Radio 4's *Today* programme, Radio 5 Live, Radio 4's *Women's Hour* and *Science Now*, GLR, BBC Local Radio, Independent Broadcasting and BBC Southern Counties Radio.

Congress was set up to provide a forum to advance physics in industry, commerce, teaching and research; to demonstrate and promulgate the benefits of physics and the contribution of physicists to learning and wealth creation; to raise the profile and influence of the Institute in the scientific and technological communities and amongst national policy makers. The increasing success of Congress testified to the Institute's vibrancy in the last decade of the millennium.

What of the future?

The Royal Charter charges the Institute 'to promote the advancement and dissemination of a knowledge of and education in the science of physics, pure and applied'.

Through its branches, groups and divisions, the Institute is striving to promote the advancement of physics in all its guises. Through its publishing company, as also through conferences and other departments in Portland Place, the Institute is working to disseminate knowledge in an effective manner.

But there is one word in that Charter—education—which will continue to be emphasised. At the end of the 1990s, there were challenges for the Institute concerned with physics education in schools and colleges in the UK. The Institute continued to invest in projects and programmes to make physics more attractive, relevant and exciting for youngsters. It will be well into the 21st century before we can measure the impact of the initiatives currently under way.

The Institute has developed in ways that could not have been predicted when pioneers like Guthrie, Glazebrook and Major Phillips had the vision to establish first the Physical Society and then the Institute. All physicists wherever they are can derive satisfaction from what has been achieved in the 125 years since 1874. The Institute of Physics is looking forward with confidence into the future.

Appendix A

BRIEF CHRONOLOGY

1873   Preliminary meeting held on 29 November to consider the formation of a physical society

1874   First meeting of the Physical Society of London on 14 February
First publication of *Proceedings of the Physical Society of London*

1905   First Annual Exhibition held in December at Imperial College of Science and Technology

1914   First Guthrie Lecture given in December by R W Wood on 'Radiation of Gas Molecules Excited by Light'

1916   Designatory letters FPSL adopted

1918   Meeting held on 27 March to discuss the formation of an institution of physics

1919   First meeting of the Board of the Institute of Physics held on 17 January

1920   Incorporation of the Institute of Physics; approval of its Memorandum and Articles of Association

1921   Inaugural meeting of the Institute held on 27 April

1922   First publication of *Journal of Scientific Instruments* by the Institute

1924   The Physical Society's jubilee celebrations (21–22 March); banquet at the Connaught Rooms in the presence of HRH The Duke of York

1927   1 Lowther Gardens becomes the Institute's headquarters; the Physical Society has offices at the same address

1932   Merger of the Physical Society of London and the Optical Society to form the Physical Society
First branch of the Institute formed in Manchester

1933   The affairs of the Physical Society and the Institute begin to become intermingled

1934   First publication of *Reports on Progress in Physics* by the Physical Society

First international physics conference (organized jointly by the Physical Society and the Royal Society) held in England in October in conjunction with a meeting of the Union of Pure and Applied Physics

1939   Evacuation of the Institute to Reading University; the Physical Society continues to operate from 1 Lowther Gardens

1945   The Institute returns to London, to 19 Albermarle Street; the Physical Society remains at 1 Lowther Gardens

1946   The Institute moves to 47 Belgrave Square
First award of the Holweck Prize

1949   First Convention of the Institute held in Buxton

1952   Establishment of the Institute's Graduateship examination

1953   The Faraday Society, the Royal Meteorological Society and the British Institute of Radiology cease to be Participating Societies of the Institute

1956   1000th Fellow of the Institute, 2000th Associate and 1000th Graduate admitted to membership

1958   First Annual Dinner of the Institute held at the Savoy Hotel

1960   The Physical Society and the Institute of Physics amalgamate as The Institute of Physics and The Physical Society
47 Belgrave Square becomes the headquarters of the combined body; publications remain at 1 Lowther Gardens

1961   First Annual Dinner of the combined body, followed by the first Annual Representatives' Meeting

1969   Formation of the European Physical Society

1970   A Royal Charter is granted to the Institute of Physics on 6 November

1971   Presidential Badge donated to the Institute by Sir James Taylor

1974   Centenary of the founding of the Physical Society of London

1975   All publishing activity relocated to Bristol

1984   Schools' Affiliation scheme inaugurated

1985   Privy Council approval for the designation 'Chartered Physicist'

1988   Launch of *Physics World*

1989   Corporate Affiliates scheme launched

1992   Launch of the *Campaign for Physics*

1993   First Annual Congress held in Brighton
Membership exceeds 20,000

1996   HM The Queen opens 76 Portland Place

1997   Launch of *Advancing Physics* (the 16–19 Physics Initiative)

Appendix B

THE PRESIDENTS

The Presidents of the Physical Society (1874–1960)

John Hall Gladstone FRS
(1874–1876)

George Carey Foster FRS
(1876–1878)

William Grylls Adams FRS
(1878–1880)

Sir William Thomson FRS
(Lord Kelvin of Largs)
(1880–1882)

Robert Bellamy Clifton FRS
(1882–1884)

Frederick Guthrie FRS
(1884–1886)

Balfour Stewart FRS
(1886–1888)

Arnold William Reinold FRS
(1888–1890)

William Edward Ayrton FRS
(1890–1892)

George Francis Fitzgerald FRS
(1892–1893)

Arthur William Rucker FRS
(1902 Sir Arthur)
(1893–1895)

William de Wiveleslie Abney FRS
(1900 Sir William)
(1895–1897)

Shelford Bidwell FRS
(1897–1899)

Oliver Joseph Lodge FRS
(1902 Sir Oliver)
(1899–1901)

Silvanus Phillips Thompson FRS
(1901–1903)

Richard Tetley Glazebrook FRS
(1917 Sir Richard)
(1903–1905)

John Henry Poynting FRS
(1905–1906)

John Perry FRS
(1906–1908)

Charles Chree FRS
(1908–1910)

Hugh Longbourne Callendar FRS
(1910–1912)

Arthur Schuster FRS
(1920 Sir Arthur)
(1912–1914)

Sir Joseph John Thomson FRS
(1914–1916)

Charles Vernon Boys FRS
(1935 Sir Charles)
(1916–1918)

Charles Herbert Lees FRS
(1918–1920)

Sir William Henry Bragg FRS
(1920–1922)

Alexander Russell FRS
(1922–1924)

Frank Edward Smith FRS
(1931 Sir Frank)
(1924–1926)

Owen Willans Richardson FRS
(1939 Sir Owen)
(1926–1928)

William Henry Eccles FRS
(1928–1930)

Sir Arthur Stanley Eddington FRS
(1930–1932)

Alexander Oliver Rankine FRS
(1932–1934)

Lord Rayleigh FRS
(Robert John Strutt)
(1934–1936)

Thomas Smith FRS
(1936–1938)

Sir Allan Ferguson
(1938–1941)

Sir Charles Galton Darwin FRS
(1941–1943)

Edward Neville da Costa Andrade FRS
(1943–1945)

David Brunt FRS
(1949 Sir David)
(1945–1947)

George Ingle Finch FRS
(1947–1949)

Sydney Chapman FRS
(1949–1950)

Leslie Fleetwood Bates FRS
(1950–1952)

**Richard Whiddington FRS
(1952–1954)**

**Harrie Stewart Wilson Massey FRS
(1960 Sir Harrie)
(1954–1956)**

**Nevill Francis Mott FRS
(1962 Sir Nevill)
(1956–1958)**

**John Ashworth Ratcliffe FRS
(1958–1960)**

The Presidents of the Institute of Physics (1920–1960)

Sir Richard Tetley Glazebrook FRS
(1920–1921)

Sir Joseph John Thomson FRS
(1921–1923)

Sir Charles Parsons FRS
(1923–1925)

Sir William Henry Bragg FRS
(1925–1927)

Sir Frank Watson Dyson FRS
(1927–1929)

William Henry Eccles FRS
(1929–1931)

Lord (Ernest) Rutherford of Nelson FRS
(1931–1933)

Sir Henry George Lyons FRS
(1933–1935)

Alfred Fowler FRS
(1935–1937)

Clifford Copland Paterson FRS
(1946 Sir Clifford)
(1937–1939)

Sir (William) Lawrence Bragg FRS
(1939–1943)

Sir Frank Edward Smith FRS
(1943–1946)

Arthur Mannering Tyndall FRS
(1946–1948)

Francis Carter Toy
(1948–1950)

William Edward Curtis FRS
(1950–1952)

Charles Sykes FRS
(1952–1954)

Sir John Douglas Cockroft FRS
(1954–1956)

Oliver William Humphreys
(1956–1958)

Sir George Paget Thomson FRS
(1958–1960)

The Presidents of the Institute and the Physical Society (1960–1970)

Sir John Douglas Cockroft FRS
(1960–1962)

Sir Alan Herries Wilson FRS
(1963–1964)

Sir Gordon Brims Black
McIvor Sutherland FRS
(1964–1966)

Sir James Taylor
(1966–1968)

Malcolm Ross Gavin
(1968–1970)

The Presidents of the Institute of Physics (1970–1998)

James Woodham Menter FRS
(1970–1972)

Sir Brian Hilton Flowers FRS
(1979 Lord Flowers)
(1972–1974)

Sir (Alfred) Brian Pippard FRS
(1974–1976)

Basil John Mason FRS
(1979 Sir John)
(1976–1978)

Rendel Sebastian (Bas) Pease FRS
(1978–1980)

Sir Denys Haigh Wilkinson FRS
(1980–1982)

Sir Robert James Clayton FEng
(1982–1984)

Sir Alexander Walter (Alec) Merrison FRS
(1984–1986)

Godfrey Harry Stafford FRS
(1986–1988)

Cyril Hilsum FRS FEng
(1988–1990)

Roger John Blin-Stoyle FRS
(1990–1992)

Clive Arthur Peirson Foxell FEng
(1992–1994)

Sir Arnold Whittaker Wolfendale FRS
(1994–1996)

Brian William Manley FEng
(1996–1998)

Sir Gareth Gwyn Roberts FRS
(1998–)

OFFICERS OF THE SOCIETY AND THE INSTITUTE

Officers of the Physical Society (1874–1960)

There were two or three Honorary Secretaries. At times they were given specific areas of responsibility. Additionally, there was the appointment of Honorary Foreign Secretary.

Honorary Secretaries

1874–1875	Edmund Atkinson
1874–1888	Arnold William Reinold FRS
1875–1883	William Chandler Roberts FRS
1883–1890	Walter Baily
1888–1893	John Perry FRS
1890–1899	Thomas H Blakesley
1893–1903	Henry M Elder
1899–1906	William Watson FRS
1902–1915	William Ranson Cooper
1906–1908	William Riach Cassie
1908–1916	Samuel Walter Johnson Smith FRS
1915–1919	William Henry Eccles FRS
1916–1919	Richard Smith Willows
1919–1920	Herbert Stanley Allen FRS
1919–1924	Frank Edward Smith FRS
1920–1925	David Owen
1924–1929	Alexander Oliver Rankine FRS
1929–1937	Ezer Griffiths FRS

1929–1938   Allan Ferguson
1937–1948   Wilfred Jevons
1938–1947   James Henry Awbery
1947–1961   Harold Horace Hopkins FRS
1948–1961   Charles Gorrie Wynne FRS
1958–1961   Archibald George Peacock

Honorary Foreign Secretaries

1896–1901   Silvanus Phillips Thompson FRS
1901–1902   Hugh Longbourne Callendar FRS
1902   Richard Tetley Glazebook FRS (1920 Sir Richard)
1903–1913   Silvanus Phillips Thompson FRS
1914–1920   Richard Tetley Glazebrook FRS (1920 Sir Richard)
1920–1928   Sir Arthur Schuster FRS
1928–1944   Owen Willans Richardson FRS
1944–1960   Edward Neville da Costa Andrade FRS

Honorary Treasurers of the Physical Society

1874–1887   Edmund Atkinson
1887–1892   Arthur William Rücker FRS
1892–1900   Edmund Atkinson
1900–1910   Hugh Longbourne Callendar FRS
1910–1918   William Duddell FRS
1918–1925   William Ranson Cooper
1925–1935   Robert Stewart Whipple
1935–1938   Robert William Paul
1938–1946   Clifford Copland Paterson FRS (1946 Sir Clifford)
1946–1950   Herman Shaw
1950–1958   Albert John Philpot
1958–1961   Donald Arthur Wright
1961   James Taylor (1966 Sir James)

Officers of the Institute of Physics

Honorary Secretaries

1920–1926   Alfred William Porter FRS
1926–1932   Alexander Oliver Rankine FRS
1932–1946   James Arnold Crowther FRS
1946–1956   Bernard Phineas Dudding

1956–1959   Francis Arthur Vick (1973 Sir Francis)
1960–1966   Charles Gorrie Wynne FRS
1966–1976   Robert Press
1976–1984   Edwin Roland Dobbs
1984–1994   Derek Humphrey Martin
1994–   Eric Jakeman FRS

Honorary Treasurers

1920–1925   Sir Robert Abbott Hadfield FRS
1925–1946   Charles Edmund Stanley Phillips
1946–1952   Edward Roy Davies
1952–1956   Stanley Whitehead
1956–1966   James Taylor (1966 Sir James)
1966–1972   Peter Thomson Menzies (1972 Sir Peter)
1972–1977   Hyman Rose
1977–1980   John Mark Anthony Lenihan
1980–1982   William Michael Lomer
1982–1988   James Moncur Valentine
1988–1999   John Logan Lewis

Secretaries/Chief Executives

1920–1925   Frederick Solomon Spiers
1926–1927   Thomas Martin
1927–1930   John James Hedges
1930–1965   Herbert Raphael Lang
1966–1990   Louis Cohen
1990–   Alun Denry Wynn Jones

Appendix D

NOBEL LAUREATES IN MEMBERSHIP

Physics

1904	Lord Rayleigh *President of the Physical Society*
1906	Sir Joseph John Thomson *President of the Physical Society and President of the Institute*
1915	Sir William Henry Bragg *President of the Physical Society and President of the Institute*
1915	Sir William Lawrence Bragg *President of the Institute*
1921	Albert Einstein *Honorary Fellow of the Physical Society*
1924	Karle Manne Georg Siegbahn *Honorary Fellow of the Institute*
1927	Charles Thomas Rees Wilson *Honorary Fellow of the Institute*
1928	Sir Owen Willans Richardson *President of the Physical Society*
1929	Prince Louis-Victor de Broglie *Honorary Fellow of the Institute*
1930	Sir Chandrasekhara Venkata Raman *Fellow of the Institute*
1932	Werner Heisenberg *Honorary Fellow of the Institute*
1933	Paul Adrien Maurice Dirac *Honorary Fellow of the Institute*
1935	Sir James Chadwick *Honorary Fellow of the Institute*

1937   Sir George Paget Thomson
       President of the Institute
1939   Ernest Orlando Lawrence
       Honorary Fellow of the Physical Society
1947   Sir Edward Victor Appleton
       Fellow of the Physical Society
1948   Lord Patrick Maynard Stuart Blackett
       Fellow of the Physical Society
1950   Cecil Frank Powell
       Honorary Fellow of the Institute
1951   John Douglas Cockroft
       President and Honorary Fellow of the Institute
1951   Ernest Thomas Sinton Walton
       Honorary Fellow of the Institute
1954   Max Born
       Fellow of the Physical Society
1955   Willis Eugene Lamb
       Honorary Fellow of the Institute
1956   John Bardeen
       Honorary Fellow of the Institute
1957   Chen Ning Yang
       Fellow of the Institute
1961   Rudolf Ludwig Mössbauer
       Honorary Fellow of the Institute
1962   Lev Davidovich Landau
       Honorary Fellow of the Institute
1970   Louis Néel
       Honorary Fellow of the Institute
1971   Dennis Gabor
       Fellow of the Institute
1972   John Bardeen
       Honorary Fellow of the Institute
1973   Brian D Josephson
       Fellow of the Institute
1974   Antony Hewish
       Fellow of the Institute
1977   Philip W Anderson
       Honorary Fellow of the Institute
1977   Sir Nevill Francis Mott
       President of the Physical Society

1978   Pyotr Leonidovich Kapitsa
Fellow of the Institute
1979   Abdus Salam
Fellow of the Physical Society
1981   Arthur L Schawlow
Fellow of the Institute
1983   Subramanyan Chandrasekhar
Honorary Fellow of the Institute
1987   Alexander K Müller
Fellow of the Institute
1988   Leon M Lederman
Fellow of the Institute
1990   Jerome I Friedman
Fellow of the Institute
1991   Pierre-Gilles de Gennes
Fellow of the Institute
1993   Russell A Hulse
Fellow of the Institute
1995   Martin L Perl
Fellow of the Institute
1996   David M Lee
Fellow of the Institute
1996   Robert C Richardson
Fellow of the Institute
1997   Steven Chu
Fellow of the Institute
1997   Claude Cohen-Tannoudji
Fellow of the Institute

Chemistry

1908   Lord Ernest Victor Buckley Rutherford
President of the Institute
1936   Petrus (Peter) Josephus Wilhelmus Debye
Honorary Fellow of the Institute
1944   Otto Hahn
Honorary Fellow of the Institute

Peace

1995   Joseph Rotblat
        Fellow of the Institute

Appendix E

CORPORATE AFFILIATES

1989	AEA Technology
1995	Alcatel Submarine Networks
1989	Alcan International
1998	AWE plc
1997	Bookham Technology
1992	BICC Cables Ltd
1989	British Gas
1989	British Nuclear Fuels Ltd
1996	British Petroleum
1997	BT Laboratories
1996	DERA (Defence Evaluation & Research Agency)
1997	GEC-Marconi
1996	Hewlett-Packard
1989	ICI
1989	Johnson Matthey
1993	Magnox Electric plc (formerly Nuclear Electric)
1997	Marconi Communications Ltd (formerly GPT)
1996	Mitel Semiconductors (formerly GEC-Plessey Semiconductors)
1989	NAG
1989	National Grid Company (formerly a division of CEGB)
1990	National Physical Laboratory
1989	National Power (formerly a division of CEGB)
1992	Nortel (formerly Northern Telecom/BNR)
1992	Nycomed-Amersham (formerly Amersham International)
1992	Oxford Instruments
1994	Particle Physics & Astronomy Research Council
1989	Pilkington plc (formerly Pilkington Technology)
1989	Post Office Research
1989	PowerGen plc (formerly a division of CEGB)
1989	Thorn-EMI

1989   Transport & Road Research Laboratory
1996   UKAEA Fusion
1997   Willett International Ltd

BRANCHES, GROUPS AND DIVISIONS

Branches

UK and Ireland

Branch	Date of formation	Membership in 1998
Irish	1964	761
Lancashire & Cumbria	1968 (*see note 1*)	396
Liverpool & North Wales	1955	668
London & South Eastern	(*see note 2*)	5692
Manchester & District	1932	1004
Midland	1935	2232
North Eastern	1949	543
Scottish	1944	1227
South Central	(*see note 2*)	1701
South Wales	1945	377
South Western	1956	1171
Yorkshire	1952	845

Notes

1 Originally the Lancaster Sub-Branch formed in 1967 as part of the Manchester & District Branch but became an independent Branch in 1968.
2 These two Branches grew from the London & Home Counties Branch which was established in 1936. In 1984 the Branch was

renamed the London & South Eastern Branch when its geographical area was divided and the South Central Branch (operating as the Guildford Sub-Branch since 1973) was formed to assume responsibility for the Surrey, Hampshire and West Sussex areas.

Overseas

The first overseas branch of the Institute was the Australian Branch which traces its origins back to 1928. By 1963, the Branch attained autonomy as 'The Australian Institute of Physics' at which stage its Council passed a motion of gratitude to the Institute of Physics in London and donated a gift as a reminder of the filial relationship between the new Institute and 'The Institute of Physics and the Physical Society'. At this stage a separate New Zealand Branch was formed, its Constitution and Rules being revised in 1972. In 1982 it attained autonomy as 'The New Zealand Institute of Physics'.

An Indian Branch, formerly functioning as a local Committee to help deal with membership applications, was established in 1934 but by 1956 the Branch was dissolved because it no longer seemed to have a purpose.

In 1956, a Branch was established in Malaya but in 1973 the Branch had to be dissolved because it was at that stage illegal both in Malaya and Singapore to operate a Branch of an organisation based outside the national frontiers of those countries. There is now in existence a new Malaysian Institute of Physics (Institut Fizik Malaysia).

Groups

Group	Date of formation	Membership in 1998
Advanced Systems	1944	72
Carbon (*see note 1*)	1965	65
Chemical Physics	1996	54
Combustion Physics	1972	16
Computational Physics	1969	773
Consultancy	1992	284
Education	1949	1140

Electron Microscopy and Analysis (*see note 2*)	1946	36
Engineering Physics (*see note 3*)	1996	266
Environmental Physics	1990	493
Gas Phase Collisions	1997	27
High Energy Particle Physics	1985	472
History of Physics	1984	380
IT, Electronics & Communications (*see note 4*)	1942	868
Instrument Science & Technology	1983	684
Ion & Plasma Surface Interactions (*see note 5*)	1970	175
Low Temperature	1945	303
Magnetism	1965	424
Materials & Characterization (*see note 6*)	1941	543
Mathematical Physics	1982	544
Medical Physics	1995	385
Neutron Scattering (*see note 7*)	1972	158
Nuclear Physics (*see note 8*)	1973	372
Optical (*see note 9*)	1899	1034
Physical Acoustics (*see note 10*)	1984	191
Physical Crystallography (*see note 11*)	1943	197
Plasma Physics	1968	284
Polymer Physics (*see note 12*)	1970	398
Printing, Packaging & Papermaking	1985	126
Quantum Electronics	1972	543
Semiconductor Physics	1982	854
Spectroscopy (*see note 13*)	1946	383
Static Electrification	1967	90
Stress & Vibration (*see note 14*)	1947	156
Superconductivity	1996	176
Theory of Condensed Matter	1995	181
Thin Films & Surfaces (*see note 15*)	1969	727
Total Quality Management	1993	152
Tribology	1980	136
Vacuum	1964	236
Women in Physics	1994	318

The Physical Society had four groups: Acoustics, Colour, Low Temperature and Optical. When the Society amalgamated with the Institute, the Colour Group opted for autonomy and separated from the amalgamated body.

Notes

1   Originally known up to 1967 as the Carbon & Graphite Group. This is a joint Group with the Royal Chemical Society and the Society of Chemistry Industry.
2   Originally known up to 1962 as the Electron Microscopy Group.
3   Created as a Division in 1970 but dissolved as such in 1991.
4   Originally known as the Electronics Group but renamed in 1996.
5   Originally known as the Atomic Collisions in Solids Group.
6   Originally known as the Industrial Radiology Group then (1953) as the Non-Destructive Testing Group, then (1961) as the Materials & Testing Group. The Group was further renamed the Materials & Characterization Group in 1994.
7   Joint Group with the Royal Society of Chemistry.
8   Originally known as the Particle–Nuclei Interactions Group then (1976) as the Nuclear Interactions Group.
9   The Optical Group had initially been the Optical Society but joined forces with the Physical Society and became the second of the Physical Society's Groups.
10 Originally a Group of the Physical Society but disbanded as a Group of the Institute upon the formation of the Institute of Acoustics (IOA) in 1974. However, since members of the Institute of Physics (IOP) in the IOA found that the new body did not adequately cater for their interests it was decided that the two bodies should form a new joint Group known as the Physical Acoustics Group.
11 Originally known as the X-ray Analysis Group. This is a joint Group with the British Crystallography Association.
12 Joint Group with the Royal Society of Chemistry and the Institute of Materials.
13 Originally known as the Industrial Spectroscopy Group which was then (1957) renamed as the Applied Spectroscopy Group before assuming its present title.
14 Originally known as the Stress Analysis Group which was renamed in 1994.

15   The Thin Films and Surfaces Group has an interesting history.
     In the 1960s the Electronics Group, the Electron Microscopy and
     Analysis Group and the Vacuum Group collaborated on a series
     of conferences on the topic of thin films and surfaces. Such was
     the success of those conferences that it was thought a new group
     devoted to thin films and surfaces should be formed. The prime
     leaders in this venture were the first two of the groups and the
     Vacuum Group stood out against the proposal on the grounds that
     it would infringe directly on its interests, which were then almost
     solely concerned with surface deposition under UHV. At a meeting
     during the third joint conference to discuss the formation of the new
     group, battle lines were drawn, tempers and emotions ran high, but
     the proponents carried the day. It is only in retrospect that the
     wisdom behind the formation of the new group was appreciated:
     this was demonstrated in the early 1990s when the Vacuum Group,
     the Thin Films and Surfaces Group, and the Ion and Plasma
     Surface Interactions Group were instrumental in the formation of
     the Surface Science and Technology Division was formed.
16   A Space Science & Astrophysics Group was formed in 1969 but
     this was dissolved in 1971.

Divisions

In 1998, the divisions were:

Applied Optics
Atomic, Molecular, Optical and Plasma Physics
Condensed Matter and Material Physics (Solid State Physics prior to
1990)
Nuclear and Particle Physics
Surface Science and Technology

When the Institute and the Society were amalgamated, the Council and
various committees were restructured. The standing committee dealing
with meetings and conferences was the Meetings Committee and it had
two sub-committees covering the (then) broad areas of physics, Nuclear
Physics and Solid State Physics. By 1966 the Atomic and Molecular
Physics sub-committee was established. The primary concern of each
sub-committee was its eponymous annual conference.

By the mid-1980s the sub-committees had ceased, in all but name, to be sub-committees of the Meetings Committee. They were operating in a similar manner to the groups. It was decided to change their status from sub-committees to divisions to which groups could be affiliated. Groups could also be involved with more than one division, thus reflecting the ever diversifying nature of physics.

The Applied Optics Division was formed differently from the other divisions in response to incursions by American optical societies into Europe, in particular SPIE (The Society for Photooptical Instrumentation Engineers). This action stimulated the European optical societies into attempting some coordination of their activities. However, a wide variety of organisations and structures covering an even wider range of optical topics had to be drawn together, and the Institute's Optical Group, although one of the largest groups, did not adequately span the UK interests. It was decided, therefore, that an Applied Optics Division should be set up. This included amongst its committee members not only representatives from other groups such as Quantum Optics and Spectroscopy, but also from other UK optical organisations such as SIRA and from the IEE.

Appendix G

AWARDS AND PRIZES

The following medals and prizes are now within the gift of the Council of the Institute of Physics.

Premier awards

The *Glazebrook Medal and Prize* is awarded annually for outstanding contributions in the organisation, utilisation or application of science.

The *Guthrie Medal and Prize* is awarded annually to a physicist of international reputation for his/her contributions to physics.

 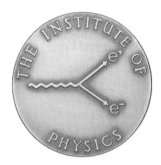

The *Paul Dirac Medal and Prize* is awarded annually for outstanding contributions to theoretical (including mathematical and computational) physics.

Principal awards

The *Holweck Medal and Prize* is awarded for distinguished work in experimental physics, or theoretical physics if closely related to experimental work. The award is made by the Council of the Institute to a French physicist in odd-dated years and by the Council of the Société Française de Physique to a British or Irish one in even-dated years.

The *Max Born Medal and Prize* is awarded for outstanding contributions to physics. The award is made alternately by the Councils of the Institute and the German Physical Society to a physicist selected from a list of nominees submitted by the other. In odd-dated years the award is made to a British or Irish physicist and is presented in Germany; in

even-dated years the award is to a German physicist and is presented in England.

The *Harry Massey Medal and Prize* is awarded for contributions to physics or its applications; normally the recipient will have, or have had, association with Australia or physics in Australia. In even-dated years the award is made by the Council of the Institute and in odd-dated years by the Australian Institute of Physics, in each case to a physicist selected from a list of nominees submitted by the other.

Senior awards

The *Charles Vernon Boys Medal and Prize* is awarded annually for distinguished research in experimental physics; the recipient is normally not more than 35 years of age at the time of the award.

The *Bragg Medal and Prize* is awarded annually for distinguished contributions to the teaching of physics.

The *Charles Chree Medal and Prize* is awarded in odd-dated years for distinguished research in one or more of the following branches of knowledge: terrestrial magnetism, atmospheric electricity and related subjects, such as other aspects of geophysics comprising the earth, oceans, atmosphere and solar–terrestrial problems.

The *Duddell Medal and Prize* is awarded annually to a person who has contributed to the advancement of knowledge by the invention or design of scientific instruments or by the discovery of materials used in their construction, or has made outstanding contributions to the application of physics.

The *Kelvin Medal and Prize* is awarded annually for outstanding contributions to the public understanding of physics.

The *Maxwell Medal and Prize* is awarded annually for outstanding contributions to theoretical physics made in the 10 years preceding the date of the award; the recipient is normally not more than 35 years old in the year of the award.

The *Paterson Medal and Prize* is awarded annually for outstanding contributions to the development, invention or discovery of new systems, processes or devices which show the successful commercial exploitation of physics; the recipient is normally not more than 40 years old in the year of the award.

The *Rutherford Medal and Prize* is awarded in even-dated years for contributions to nuclear physics, elementary particle physics or nuclear technology.

The *Thomas Young Medal and Prize* is awarded in odd-dated years for distinguished work on optics.

Other awards

The *Simon Memorial Prize* is awarded approximately every three years by the Committee of the Low Temperature Group for distinguished work in experimental or theoretical low temperature physics.

The Charles Vernon Boys Medal and Prize.

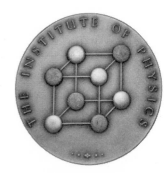

The Bragg Medal and Prize.

The Charles Chree Medal and Prize.

The Duddell Medal and Prize.

The Kelvin Medal and Prize.

The Maxwell Medal and Prize.

The Paterson Medal and Prize.

The Rutherford Medal and Prize.

The Thomas Young Medal and Prize.

The *Teachers of Physics Awards* are made by the Committee of the Education Group to school teachers in secondary education.

The *Teachers of Primary Science Awards* are made by the Committee of the Education Group.

Honorary Fellowship

Bylaw 4 of the *Charter and Bylaws* defines Honorary Fellows as follows:

> 'Distinguished persons intimately connected with physics or a science allied thereto whom the Institute especially desires to honour for exceptionally important services in connexion therewith, and any distinguished person whom the Institute may desire to honour for service to the Institute or whose association therewith is of benefit to the Institute, shall be elegible to become Honorary Fellows of the Institute. The total number of Honorary Fellows for the time being shall not exceed twenty.'

A candidate for Honorary Fellowship shall be nominated only by the Council.

Appendix H

CHANGES IN MEMBERSHIP CLASSES

Before the date of Charter	At the date of Charter

The corporate classes of membership

Honorary Fellow (HonFInstP)	Honorary Fellow (HonFInstP)
Fellow (FInstP)	Fellow (FInstP)
Associate (AInstP)	Member (MInstP)
Fellow of the Physical Society *	Fellow of the former

The non-corporate classes of membership

Graduate Member (GradInstP)	Associate Member * (*see note 2*)
Licentiate (LInstP)	Associate *
Student *	Student *

Other

Subscriber *	Subscriber* (*see note 3*)

The above is the situation which applied up to and at the date of Royal Charter (6 November 1970).

Notes

1   An asterisk denotes that no right was granted by Privy Council to use designatory letters.
2   The Associate class was merged into the Associate Member class in 1982.   In 1994, the Graduate Member class was restored for those with an academic qualification acceptable for corporate membership. Graduate Members were then given the right to use

the designatory letters 'GradInstP'. Others who did not possess an academic qualification acceptable for corporate membership were then transferred into the restored class of Associate but were not given the right to use designatory letters.

3   Subscribership (renamed 'Affiliate' in 1992), strictly speaking, is not a membership class.

4   From January 1985, all corporate members were given the right to use the title 'Chartered Physicist' and the designation 'CPhys'.